* HIGGS BOSON *

* ORIGIN OF BIG BANG *

THE UNIVERSE

SPACE

AND BEYOND

* Basic Physics *   * Particle Physics *

* Tetra and Penta Quarks *

* Relativity *   * Cosmology *

* Origin of Dark Matter *

* Beyond Our Universe *

* Inter-Universe Travel *

* Before and After Our Universe *

7TH EDITION

OCTOBER 13, 2016

JOHN GILBERT BEAN

JohnGilbertBean.wordpress.com

Library of Congress Card Catalog Number – 2011902227

Publisher's Cataloging-in-Publication

Bean, John Gilbert, 1933 –

    THE UNIVERSE, SPACE, AND BEYOND   / Maureen Bean – ed.

    p. cm.

    LCCN: 2011902227

    Includes index

    1. General Physics   2. Particle Physics)

    3. Relativity (Physics)   4. Cosmology – Popular Works

    I. Title

ISBN-13: 978-1497418646
ISBN-10: 149741864X

Published by: John Gilbert Bean Publishing
        SAN: 253-0287

        951-265-1124
        Send comments to: JBeanMBean@msn.com

1st EDITION February 13, 2011   2nd EDITION April 13, 2013   July 13, 2013
3rd EDITION December 13, 2013   4th EDITION March 13, 2014
5th EDITION July 13, 2014   October 1, 2014
6th EDITION April 13,2015   July 13, 2015   October 13,2015   November 13, 2015
7th EDITION Jun 13, 2016   October 13, 2016

#CONTENTS

NOTE: For access on your viewer using its "Find" capability, each **section** number is preceded by #. For example, to go to **Section 2**, enter #2.

PREFACE
DEDICATION
ABOUT THE AUTHOR

PART I. <u>GENERAL INFORMATION ABOUT EARTH AND UNIVERSE.</u>

1. TIMELINE , IMPORTANT CONSTANTS, AND DEFINITIONS.
1.1. TIMELINE.
1.2. IMPORTANT CONSTANTS AND DEFINITIONS.
1.2.1. EINSTEIN'S EQUATION OF THE EQUIVALENCE OF MASS m AND ENERGY E.
1.2.2. FORCE.
1.2.3. DISTANCE, WEIGHT, GRAVITY, GRAVITATIONAL FORCE, GRAVITATIONAL FIELD, AND CENTER OF MASS.
1.2.4. MASS (IN KILOGRAMS) OF VARIOUS OBJECTS.
1.2.5. WORK AND ENERGY.
1.2.6. KINETIC ENERGY.

1.2.7. POWER.
1.2.8. LINEAR AND ANGULAR MOMENTUM.
1.2.9. VELOCITY, FREQUENCY, AND WAVELENGTH OF LIGHT.
1.2.10. ENERGY (E) OF PHOTON.
1.2.11. EQUIVALENT MASS OF PHOTON.
1.2.12. ELECTROMAGNETIC SPECTRUM OF PROTONS.
1.2.13. COSMOLOGICAL UNITS.
1.2.14. AGE AND DIAMETER OF UNIVERSE.
1.2.15. ELECTRON VOLT eV, MASS AND CHARGE OF ELECTRON AND PROTON
1.2.16. COSMIC RAYS.

CONTENTS (CONTINUED)

1.2.17. PLANCK AND BIG BANG SPHERE MEASUREMENTS.

1.2.18. DIAMETER OF PROTON AND ELECTRON.

1.2.19. LENGTH, DIAMETER, AND MASS OF VIBRATING QUANTUM STRING OF ENERGY Q_e OF STRING THEORY.

1.2.20. de BROGLIE WAVELENGTH λ .

1.2.21. π

1.2.22. PRESSURE.

1.2.23. TEMPERATURE (IN DEGREES).

1.2.24. RELATIONSHIP OF HEAT TO OTHER FORMS OF ENERGY.

1.2.25. ENERGY CONVERSION LIST. (SEE ALSO SECTIONS 1.2.5 AND 1.2.6.)

2. BASIC PHYSICS OF EARTH AND UNIVERSE.

2.1. MASS AND ENERGY.

2.2. MASS AND ENERGY ARE QUANTA OF SAME THING.

2.3 FORCE, MASS, GRAVITY, AND WEIGHT.

2.3.1. FORCE.

2.3.2. WEIGHT.

2.3.3. GRAVITY AND GRAVITATIONAL WAVES.

2.3.4. UNITS OF MASS AND FORCE

2.3.5. NEWTON'S LAWS OF MOTION.

2.4. WORK.

2.5. ENERGY .

2.5.1. KINETIC ENERGY.

2.5.2. POTENTIAL ENERGY, CONSERVATIVE FORCES, AND DISSIPATIVE FORCES.

2.5.3. PRINCIPLE OF CONSERVATION OF ENERGY DETERMINES VELOCITY REQUIRED TO ESCAPE FROM EARTH.

2.6. POWER.

2.7. LINEAR MOMENTUM.

2.8. ANGULAR MOTION AND ANGULAR MOMENTUM.

2.9. TRANSFER OF HEAT ENERGY.

2.9.1. TRANSFER OF HEAT ENERGY BY CONDUCTION.

2.9.2. TRANSFER OF HEAT ENERGY BY CONVECTION.

2.9.3. TRANSFER OF HEAT ENERGY BY RADIATION.

2.9.4. CHANGE OF STATE.

2.10. PRESSURE.

2.10.1. UNITS OF MEASUREMENT OF PRESSURE.

2.10.2. EXAMPLES OF PRESSURE.

2.10.3. POSITIVE ENERGY PRESSURE INSIDE BIG BANG SPHERE AND NEGATIVE ENERGY PRESSURE OF EMPTY SPACE OUTSIDE BIG BANG SPHERE ARE KEY TO CREATING OUR UNIVERSE.

3. OVERVIEW OF WAVE MOTION AND LIGHT CHARACTERISTICS.

3.1. OSCILLATING SINUSOIDAL WAVES TRAVEL OUTWARD FROM THEIR SOURCE.

3.2. TRANSVERSE AND LONGITUDINAL WAVES.

3.3. AMPLITUDE, FREQUENCY, WAVELENGTH, PERIOD, PHASE, AND VELOCITY OF WAVES.

3.4. PHASE OF WAVE IS MEASURED IN 360 DEGREES.

3.5. WAVE VELOCITY.

3.5.1. SOUND VELOCITY.

3.5.2. ELECTROMAGNETIC WAVE VELOCITY.

3.6. FUNDAMENTAL FREQUENCY AND HARMONICS OF WAVES.

3.6.1. DOPPLER EFFECT.

3.7. REFLECTION OF WAVES.

3.8. REFRACTION OF WAVES.

3.9. INTERFERENCE - WAVES CANCEL AND REINFORCE EACH OTHER AND THEMSELVES.

3.10. COHERENT WAVES.

3.11. PERIODIC WAVES AND BEAT FREQUENCIES (NOTES).

3.12. POLARIZED LIGHT.

CONTENTS (CONTINUED)

4. LIGHT RADIATION AND SPECTRUM.

4.1. HEATING OBJECTS TO VARIOUS CONSTANT TEMPERATURES.

4.2. QUANTUM THEORY OF LIGHT: BLACKBODY RADIATION - ORIGIN OF QUANTUM MECHANICS.

4.3. PHOTOELECTRIC EFFECT.

5. ELECTROMAGNETIC RADIATION AND PHOTONS.

5.1. VELOCITY, FREQUENCY, AND WAVELENGTH OF LIGHT.

5.2.   ELECTROMAGNETIC WAVE SPECTRUM.

5.3. ENERGY (E) OF PHOTON.

5.4. EQUIVALENT MASS OF PHOTON.

5.5. DEFLECTION OF PHOTONS BY GRAVITY.

6. OUR UNIVERSE.

6.1. COSMOLOGICAL UNITS.

6.2. TIMELINE, TEMPERATURE, AND PRODUCTS OF BIG BANG THAT CREATED OUR UNIVERSE

6.2.1. AS BIG BANG COOLED, HYDROGEN WAS FORMED.

6.2.2. TYPES OF STARS – CREATION OF ELEMENTS IN THE UNIVERSE.

6.2.3. DYING STARS ARE FACTORIES THAT CREATED ALL THE ATOMS IN THE UNIVERSE FROM HYDROGEN CLOUDS.

6.3. BLACK HOLES AND QUASARS.

6.4. OUR UNIVERSE LOOKS THE SAME IN ALL DIRECTIONS.

6.5. OUR UNIVERSE IS EXPANDING AT ACCELERATING RATE.

6.6. AGE OF UNIVERSE.

6.7. COSMIC MICROWAVE BACKGROUND RADIATION (CMBR).

6.8. DIAMETER OF UNIVERSE.

6.9. COSMIC RAYS

7. WAVE MOTION AND INTERFERENCE CHANGE UNDERSTANDING OF QUANTUM PARTICLES.
7.1. COHERENT WAVES INTERFERE WITH EACH OTHER TO PRODUCE INTERFERENCE PATTERNS.
7.2. QUANTUM PARTICLES INTERFERE WITH THEMSELVES AND ARE SEEMINGLY AT TWO (OR MORE) PLACES AT ONCE.

PART II. PARTICLE PHYSICS.

8. PARTICLE CHARACTERISTICS.
8.1. GENERAL PARTICLE CHARACTERISTICS.
8.2. ADDITIONAL CHARACTERISTICS OF QUARKS.

9. ELEMENTARY PARTICLES – FERMIONS (LEPTONS AND QUARKS) AND BOSONS (38).
9.1 ELECTROMAGNETICALLY NEUTRAL LEPTONS (NEUTRINOS) (3).
9.2. ELECTROMAGNETICALLY CHARGED LEPTONS (3).
9.3. QUARKS (18).

9.4. BOSONS (14).
9.4.1. ELECTROMAGNETIC FORCE BOSON (PHOTON) (1).
9.4.2. WEAK NUCLEAR FORCE BOSONS (2).
9.4.3. STRONG NUCLEAR (COLOR) FORCE BOSONS (GLUONS) (8).
9.4.4. GRAVITATIONAL FORCE BOSONS (3).

10. COMPOSITE PARTICLES - HADRONS (MESONS, BARYONS, TETRA-QUARKS, AND PENTA-QUARKS).
10.1. EXAMPLE MESONS.
10.2. MESON LISTINGS (118).
10.3. EXAMPLE BARYONS.
10.4. BARYON LISTINGS (86).

CONTENTS (CONTINUED)

11. ELEMENTARY PARTICLE DESCRIPTIONS.

11.1 FERMIONS - DESCRIPTION.

11.2 . LEPTONS - DESCRIPTION.

11.2.1. ELECTROMAGNETICALLY NEUTRAL LEPTONS (NEUTRINOS) - DESCRIPTION

11.2.2. ELECTROMAGNETICALLY CHARGED LEPTONS - DESCRIPTION.

11.3. QUARKS – DESCRIPTION.

11.4. BOSONS – DESCRIPTION AND RELATIVE STRENGTH.

11.5.   HADRONS (MESONS AND BOSONS) - DESCRIPTION.

11.6. MESONS - DESCRIPTION.

11.6.1. MESONS ARE COLOR AND CHARGE NEUTRAL.

11.6.2. QUARK COMBINATIONS IN MESONS.

11.7. BARYONS - DESCRIPTION.

11.7.1. BARYONS ARE COLOR NEUTRAL AND HAVE HALF-INTEGER SPIN.

11.7.2. PROTONS AND NEUTRONS ARE BARYONS MADE OF THREE QUARKS – WHERE DID ALL THE MASS COME FROM?

11.7.3. QUARK COMBINATIONS IN BARYONS.

11.7.4. QUARK COMBINATIONS IN MORE COMPLEX COMBINATIONS MAY BE ORIGIN OF DARK MATTER.

12.   PARTICLE INTERACTIONS – BACKGROUND INFORMATION.

12.1. ELECTRON VOLT Ev IS USED IN PARTICLE PHYSICS FOR ENERGY = Ev, MOMENTUM = Ev/c, AND MASS = Ev/c^2.

12.2. CONSERVATION OF ENERGY, MOMENTUM, CHARGE, AND OTHER KEY PARTICLE CHARACTERISTICS APPLY TO EVERY INTERACTION (MEDIATION) BETWEEN PARTICLES.

12.2.1. RELATIVISTIC MASS AND ENERGY OF PARTICLES ARE CONSERVED IN PARTICLE INTERACTIONS.

12.2.2. OTHER KEY CHARACTERISTICS OF PARTICLES ARE THE SAME (CONSERVED) BEFORE AND AFTER INTERACTIONS.

12.3.   ELEMENTARY PARTICLES, HADRONS, ATOMS, MOLECULES, AND EVEN LARGE OBJECTS EXHIBIT BOTH PARTICLE AND de BROGLIE WAVE CHARACTERISTICS.

12.4.   UNCERTAINTY PRINCIPLE AND PROBABILITIES RULE AT THE QUANTUM LEVEL - IF SOMETHING CAN HAPPEN, IT WILL, BUT MAYBE NOT VERY OFTEN.

12.4.1. UNCERTAINTY PRINCIPLE.

12.4.2. PROBABILITY WAVES.

12.4.3. WHICH OF BILLIONS AND BILLIONS OF POSSIBLE PATHS WILL A SINGLE PARTICLE SUCH AS A PHOTON TAKE?

12.5.   UNCERTAINTY PRINCIPLE AND de BROGLIE WAVES EXPLAIN TWO EARLY MYSTERIES OF CHARGED PARTICLES IN ATOMIC AND HADRON STRUCTURES.

12.6.   IF STATES OF TWO OR MORE ENTANGLED PARTICLES ARE MANDATED OR PROHIBITED, ALLOWED STATES ARE INSTANTLY ENFORCED EVERYWHERE IN THE UNIVERSE WHEN ARBITRARILY SELECTED BY ENTANGLED PARTICLE.

12.7.   EMPTY SPACE IS NOT EMPTY.

12.7.1. EMPTY SPACE WITHIN OUR UNIVERSE IS NOT EMPTY.

12.7.2. EMPTY SPACE BEYOND OUR UNIVERSE (IF IT EXISTS) IS NOT EMPTY.

13. PARTICLE INTERACTIONS: THE HEART OF PARTICLE PHYSICS.

13.1 WAYS THAT PARTICLES INTERACT.

13.1.1. INTERACTION NOTATION

13.1.2. FEYNMAN DIAGRAMS.

13.2.   PARTICLE AND ANTI-PARTICLE INTERACTIONS.

13.3.   ELECTROMAGNETIC FORCE BOSONS (PHOTONS) MEDIATE INTERACTIONS BETWEEN CHARGED PARTICLES – COULOMB'S LAW.

13.3.1. ELECTROMAGNETIC CHARGES OF ELEMENTARY PARTICLES ARE INFINITE TO ENABLE ELECTROMAGNETIC INTERACTION.

13.3.2. PROPERTIES OF EMPTY SPACE SCREEN TRUE VALUE OF PARTICLE ELECTRIC CHARGE.

13.3.3. PHOTON SPIN ENABLES ELECTRIC CHARGES TO ATTRACT OR REPEL EACH OTHER.

13.4. STRONG NUCLEAR (COLOR) FORCE BOSONS (GLUONS) HOLD QUARKS TOGETHER.

13.4.1. STRONG NUCLEAR (COLOR) FORCE CONSISTS OF EIGHT GLUONS.

13.4.2. QUARKS INTERACT WITH THE STRONG NUCLEAR (COLOR) FORCE GLUONS.

13.4.3. HADRONS ARE PRODUCED WHEN QUARKS INTERACT WITH GLUONS.

CONTENTS (CONTINUED)

13.4.4. STRONG NUCLEAR (COLOR) FORCE GLUONS ARE INFINITE IN VALUE AND LONG RANGE IN PRINCIPLE, BUT ARE USUALLY CONFINED WITHIN A HADRON (MESON OR BARYON).

13.4.5. WHERE DID ALL THE MASS OF THE PROTON COME FROM?

13.5. THREE WEAK NUCLEAR FORCE BOSONS W^+ W^- Z^0 CONTROL PARTICLE INTERACTION AND DECAY.

13.5.1. ALL FERMIONS (LEPTONS AND QUARKS) INTERACT WITH WEAK CHARGE.

13.5.2. WEAK NUCLEAR FORCE BOSON W^+ W^- Z^0 INTERACTION AND DECAY EQUATIONS.

13.6. PARTICLE INTERACTION AND DECAY RULES.

13.7. HIGGS BOSONS AND GRAVITON BOSONS ADD MASS AND GRAVITATIONAL FORCE TO PARTICLES AND WEAK-NUCLEAR-FORCE BOSONS.

13.7.1. PROPERTIES OF MASS.

13.7.2. THERE ARE FOUR SCALAR HIGGS FIELDS IN THE FABRIC OF SPACETIME.

13.7.3. MATHEMATICAL DESCRIPTION OF FIELDS IS DERIVED FROM LAGRANGIAN FORMULATION OF ENERGY AND MOTION.

13.7.4. HIGGS FIELDS AND BOSONS CORRECT THEORETICAL PROBLEMS.

13.7.5. DESCRIPTION 0F HIGGS FIELDS.

13.7.6. HIGGS FIELDS TAKE ON SCALAR VALUES AT EVERY POINT AND CREATE GEODESICS IN CURVED SPACE.

13.7.7.   OBJECTS WITH MASS FOLLOW GEODESIC CURVES (WARPS) IN SPACETIME.

13.7.8. GRAVITY IS MORE THAN JUST AN ATTRACTIVE FORCE.

13.7.9. ALTERNATE THEORIES OF GRAVITY.

13.7.10. DETECTION OF HIGGS BOSON IN JULY 2012.

PART III. THE UNIVERSE AND BEYOND.

14. ADVANCED PARTICLE TOPICS AND SEARCHES.

14.1. EACH ELEMENTARY PARTICLE IS THEORIZED TO HAVE CORRESPONDING SUPER-SYMMETRIC PARTICLE.

14.2. DARK MATTER: 21 PERCENT OF OUR UNIVERSE; DARK ENERGY: 72 PERCENT; ORDINARY MATTER: 7 PERCENT

14.2.1. DARK MATTER (21 PERCENT OF OUR UNIVERSE).
14.2.2. COMPOSITION OF DARK MATTER AND ITS ORIGIN.

14.3. DARK ENERGY (72 PERCENT OF OUR UNIVERSE) IS CAUSING OUR UNIVERSE TO EXPAND AT EVER-INCREASING RATE.

15.   DO BOTH ANTI-PARTICLE AS WELL AS PARTICLE UNIVERSES (SUCH AS OURS EXIST)?

16. WHAT ABOUT OTHER KINDS OF UNIVERSES?

17. PUTTING IT ALL TOGETHER – WHERE AND HOW DID BIG BANG COME ABOUT BEFORE OUR UNIVERSE EXISTED?

17.1. QUANTUM STRINGS OF VIBRATING ENERGY.

17.2. FABRIC OF EMPTY SPACE THAT EXISTED BEFORE BIG BANG CREATED VIRTUAL PARTICLES AND VIRTUAL ANTI-PARTICLES – THESE LED TO SPHERE OF BIG BANG.

17.3. SPHERE OF VIRTUAL LEPTONS, QUARKS, GLUONS, AND BARYONS MAY NOT BE ABLE TO OBTAIN HIGH ENOUGH PRESSURE, TEMPERATURE, AND DENSITY TO CAUSE BIG BANG.

CONTENTS (CONTINUED)

17.4. COMBINING OF VIRTUAL PARTICLES TO CREATE BIG BANG IS DETERMINED BY PROBABILITIES.

17.5. BIG BANG WAS CREATED BY QUANTUM VIBRATING STRINGS.

17.6. AS TEMPERATURE AND PRESSURE INCREASE IN BIG BANG SPHERE, BIG BANG OCCURS.

17.7. HOW MUCH NET ENERGY DOES IT TAKE IN BIG BANG TAKE TO MAKE OUR UNIVERSE?

17.8. VIBRATING STRINGS INITIATED BIG BANG WITH LESS PRESSURE, TEMPERATURE, AND ENERGY THAN EXPECTED.

17.9. ENERGY BORROWED FROM EMPTY SPACE TO CREATE UNIVERSE MUST BE REPAID.

17.10. PUTTING EVERYTHING TOGETHER TO FROM BEGINNING TO END OF OUR UNIVERSE.

17.11. QUANTUM WAVES AND STRINGS OF VIBRATING ENERGY ARE ESSENTIAL FOR A "BIG BANG" TO CREATE A UNIVERSE.

17.12. BAD NEWS FOR TRAVEL BEYOND OUR UNIVERSE.

17.13. OUR UNIVERSE WILL CONTINUE TO EXPAND DUE TO DARK ENERGY.

18. LONG BEFORE AND LONG AFTER THE BIG BANG: INFINITE EMPTY SPACE.

19. UNDERSTANDING NATURE, OUR UNIVERSE, AND BEYOND OUR UNIVERSE USING SCIENTIFIC METHOD.

PART IV. OVERVIEW OF MATHEMATICS OF QUANTUM THEORY AND RELATIVITY

20. SCHRÖDINGER'S EQUATION.

20.1. PROBABILISTIC CHARACTERISTICS OF WAVE-PARTICLE QUANTUM WORLD.

20.2. CONSERVATION CHARACTERISTICS OF QUANTUM WORLD.

20.3 EXPANDING SCHRÖDINGER'S EQUATION TO RELATIVISTIC PARTICLE INTERACTIONS.

21. SPECIAL RELATIVITY

21.1. OVERVIEW OF SPECIAL AND GENERAL RELATIVITY – ENERGY WARPS SPACETIME.

21.1.1. KINETIC ENERGY WARPS SPACE AND TIME (SPACETIME) COLLINEARLY.

21.1.2. POTENTIAL ENERGY WARPS (CURVES) SPACE AND TIME (SPACETIME).

21.1.3. BEHAVIOR OF LIGHT IS DIFFERENT THAN THAT OF OTHER TYPES OF WAVES.

21.1.4. ELECTROMAGNETIC WAVES PROPAGATE FROM CENTRAL POINT.

21.2. LORENTZ EQUATIONS.

21.3. LENGTH CONTRACTION AND TIME DILATION.

21.3.1. LENGTH CONTRACTION

21.3.2. TIME DILATION.

21.2. REVIEW OF CONSTANT RELATIVE MOTION.

21.3. SPECIAL RELATIVITY MATHEMATICS.

21.4. LORENTZ EQUATIONS LEAD TO THE EQUATIONS FOR THE PROPAGATION OF LIGHT.

21.5. HYPERBOLIC FUNCTIONS OF LORENTZ EQUATIONS.

21.6. COORDINATE TRANSFORMATION USING LINEAR CARTESIAN COORDINATES.

21.7. CARTESIAN ORTHOGONAL COORDINATES OF LORENTZ EQUATIONS.

CONTENTS (CONTINUED)

21.8. VECTOR-SPACETIME GRAPHS AND COORDINATE ROTATION.

21.9. EXAMPLE OF COORDINATE TRANSFORMATION.

21.10. A QUESTION OF SIMULTANEITY: WHEN WAS THE SHOT FIRED IN NEW YORK CITY?

21.11. IS "T" or "t" A FOURTH DIMENSION?

21.12.   A TIME MACHINE (THE TWIN PARADOX).
21.12.1. THE TRAVELING TWIN IS YOUNGER DUE TO TIME DILATION.
21.12.2. TRAVELING TWIN IS YOUNGER DUE TO DISTANCE CONTRACTION.
21.12.3. TRAVELING TWIN IS YOUNGER DUE TO ACCELERATION "LOCKING- IN" NEW (SLOWER) CLOCK RATES.
21.12.4. ACCELERATION AND DECELERATION OF TRAVELER'S CLOCK TAKE TRAVELER'S CLOCK (AND TRAVELER WITH IT) TO NEW (SLOWER) CLOCK RATE.
21.13. EQUATIONS OF PHYSICS MUST BE MODIFIED TO INCORPORATE SPECIAL RELATIVITY EFFECTS.

22. GENERAL RELATIVITY.

22.1. EINSTEIN'S SPECIAL RELATIVITY AND GENERAL RELATIVELY ARE VERY DIFFERENT.

22.2.   QUANTITATIVE MATHEMATICAL TREATMENT OF GENERAL RELATIVITY.
22.3. MASS AND ENERGY CAUSE SPACETIME TO CURVE (WARP).
22.3.1. ENERGY IN SPACETIME IS EQUIVALENT TO MASS.
22.3.2.   SPACETIME CURVE IS GEODESIC.
22.4.   ACCELERATION BY GRAVITY AND OTHER MEANS.
22.5. CURVING (WARPING) OF SPACE TIME BY MASS AND ENERGY PLACES OTHERWISE UNACCELERATED OBJECTS IN GEODESIC ORBITS WHERE THEY FREEFALL.

22.6. EFFECT OF GRAVITY AND ACCELERATION ON CLOCKS (TIME) AND DISTANCE.

22.7. GRAVITY AND ACCELERATION CREATE *G*EODESICS THAT PLACE OBJECTS IN FREE-FALL GEODESIC ORBITS.
22.7.1. FRICTIONLESS MERRY-GO-ROUND GEODESIC ON EARTH.
22.7.2. FRICTIONLESS MERRY-GO-ROUND GEODESIC IN OUTER SPACE
22.7.3. WEIGHT ON STRING IN GEODESIC ORBIT IN OUTER SPACE.
22.7.4. WEIGHT IN GEODESIC AROUND MASSIVE OBJECT.

22.8. METRICS ARE VERY IMPORTANT PART OF GENERAL RELATIVITY.

22.9. TRANSITIONING FROM SPECIAL TO GENERAL RELATIVITY.
22.10. GENERALIZING COORDINATES.
22.11. USING TENSORS TO DESCRIBE MOTION IN CURVED SPACETIME.

22.12. EINSTEIN'S GENERAL RELATIVITY EQUATIONS.

22.13. COMPARISON OF EINSTEIN'S EQUATIONS WITH NEWTON'S EQUATIONS.
22.13.1. PLANET MOVING AROUND THE SUN - THE PERIHELION OF MERCURY.
22.13.2. DEFLECTION OF LIGHT RAY BY MASSIVE OBJECT.

EPILOGUE.

BIBLIOGRAPHY.

INDEX

#PREFACE.

This book is about Space and time before, during, and after our Universe and its galaxies, stars, planets, moons, and other objects therein. The term "Spacetime" is used to show the close correlation of space and time and that time is essentially a fourth dimension.

This book has been published in several editions. Each new edition entirely replaces the prior edition(s). The subsequent editions significantly expanded content in many areas of basic physics, particle physics, relativity, cosmology, and Big Bang. Subsequent editions also clarified, reorganized, expanded, and included the latest available technical information.

The topics covered are explained with emphasis on qualitative and physical descriptions so as to be easily understood by **all levels of readers**. For those who wish to progress to more detailed mathematics, the descriptive material will greatly facilitate the transition. **Section 22** and the **Bibliography** list several mathematically-oriented texts.

Physics has come a long way in the last hundred years or so since 1905 when Albert Einstein revolutionized the scientific world with his famous equation Energy E equals Mass M multiplied by the square of the velocity c of Light (**$E = Mc^2$**) stating the equivalence of mass and energy. His insight that a little bit of mass creates a lot of energy led to the atomic bomb among many other developments. Also in 1905, his thinking led to his ***Special Theory of Relativity*** explaining the fundamental nature of the velocity of light and its relationship to space. Einstein's revolutionary thinking continued with his ***General Theory of Relativity*** in 1915 expanding our understanding of acceleration, gravity, mass, and energy, space, and time.

These ingenious breakthroughs have been followed by arguably equally important breakthroughs by many other physicists and cosmologists that enable understanding and explaining our Universe. One of the key discoveries was made by Edwin Hubble in 1928 that the Universe is not static but expanding and galaxies are all moving away from each other.

Cosmology exploded with the advent of advanced telescopes, optical technology, satellites, space probes, and the discovery that our Universe is expanding at an ever accelerating rate.

Using the astounding capabilities of the recent Hubble Telescope (named for Edwin Hubble) and the mathematic of Einstein's General Relativity, Physicists and Cosmologists have postulated that the rapid inflation and expansion of our infant Universe in its first fraction of a second after the Big Bang caused its formation and evolution. Cosmological theories and research of the Universe continue to multiply in quantity, accuracy, and detail.   Their main conclusions relevant here are:

- The Universe consists of many billions of Galaxies.
- Each Galaxy contains billions of stars.

- The Universe is expanding at an accelerating rate.
- Galaxies are not expanding but are moving apart from each other at an accelerating rate.
- Space between Galaxies is expanding and creating Space.

- Space and time did not exist until our Universe was created.

Taking the above information and extrapolating back in time, the conclusions are:

- Our Universe was created by a Big Bang at a tiny point.
- Everything in our Universe is a descendent from that point.
- Time and Space began at the Big Bang.
- Every Galaxy is at the center of our Universe including our Galaxy, the Milky Way.

Despite these major advances, there are still mysteries about our Universe and Space beyond:

- Where did the Big Bang happen?

-How did the Big Bang happen?

- What and where was there before the Big Bang?

How will our Universe end?

- Are there other Universes?

This book provides possible explanations of these questions.

PREFACE (CONTINUED)

The main purpose of this book is to provide the basic science necessary to understand our Earth and how it works: BASIC PHYSICS AND PARTICLE PHYSICS. From this foundation, the more complex issues are discussed and will provide a transition for advanced study.

This book is meant to be a general discussion of its subjects. It is not meant to be a research or historical paper. Nevertheless, I have been diligent to improve its accuracy in subsequent revisions and editions. The state-of-the-art in many of the applicable areas is highly speculative, untested, and unproven. In these areas I have felt comfortable with educated speculation and conjecture to provide "place-holders" until research and experiment progress.

One of my goals in life is to understand scientifically where everything came from and where it is going. This book enabled me to clarify my knowledge toward fulfilling my goal. I hope the readers will enjoy it as much as I enjoyed writing it.

This is the greatest time ever to be a physicist with enormous revolutions and breakthroughs happening in particle physics, string theory, and cosmology. Physicists look forward to each day bringing another research paper or announcement. Each day may bring an important discovery from the new Large Hadron Collider at CERN that is recently up and running. This accelerator collides (smashes together) protons and other particles traveling near the velocity of light at energy of upwards of 24 trillion eV. So far, it has detected the Higgs boson and will continue to find many as yet unseen and undetected particles. Those at the forefront of string theory continue to search for the Holy Grail that has eluded Einstein and all others, "A Unified Theory of Everything." Expectations are high that they will be successful.

Maybe physicists will find there is something beyond "Everything." Maybe the required calculations are too enormous to carry out. Maybe the complexity is beyond comprehension without an Einstein to point in a new, more-productive direction.

Perhaps the physics and make-up of each of an infinite number of possible universes happen by chance, are all radically different, and are impossible to ever comprehend. Hopefully though, as I believe, if you understand one universe, you understand them all, and the physics and physical laws are the same for all. That's the simplest explanation – and, "the simplest explanation is usually the right one."

Every physicist especially those of my age wants to jump into a spaceship, roar off by accelerating to near speed-of-light velocity to slow down their personal on-board clock (as compared to the reference clocks at home), and come back in 1000 Earth years or so to see how it all turned out.

John Gilbert Bean
JohnGilbertBean.wordpress.com

Send comments to JBeanMBean@msn.com
THE UNIVERSE, SPACE, AND BEYOND

#DEDICATION.

This work is dedicated to my wonderful wife, Maureen, who by her awesome goodness and talents gave me a taste of Paradise and the will to accomplish important things.
JGB

#ABOUT THE AUTHOR.

John Gilbert Bean has a BS Degree in Physics from the University of Arizona and graduate study in Physics and Electrical Engineering at the University of Arizona and California State University at Northridge (formerly San Fernando Valley State College).

As well as writing, his career spanned Engineering, Engineering Management, and Program Management responsible for design, development and integration of advanced systems at RCA, Hughes Aircraft, and Northrop Grumman.

His career and interests enabled him to work at the forefront of technology in the rapidly changing relevant science and mathematics of the origin and destiny of our Universe and Space beyond. He has also written: ***Communicating Successfully with Everyone.***

#PART I. <u>GENERAL INFORMATION ABOUT EARTH AND UNIVERSE.</u>

#1. <u>TIMELINE, IMPORTANT CONSTANTS, AND DEFINITIONS.</u>

#1.1. TIMELINE.

- Number of Universes in existence
    At least 1, probably many more
- Our Universe created
    13.8 billion years ago
- Number of galaxies in our Universe
    150 billion (probably more)
- Number of stars in our Universe
    30 thousand billion, billion (300 sextillion) (probably more)
- Earth and Solar System created
    4.7 billion years ago
- Intense bombardment of Earth by meteors ended
    4.5 billion years ago
- First single-cell life on Earth
    4 billion years ago
- First multi-cell life on Earth (fractals)
    580 million years ago
- First life on earth that could move around
    550 million years ago
- First mammals and dinosaurs
    230 million years ago
- Dinosaurs dominate all land inhabitants
    200 million years ago
- Dinosaurs and many other species became extinct when large meteor crashed into Earth
    65 million years ago
- Early human ancestors left trees and walked upright with stiff arched feet
    6 million years ago
- Early human ancestors knap (chip) stones to create tools with sharp points and edges
    3.3 million years ago [Nature May, 2015]

- Early human ancestors lost most of their body hair

    1 million years ago

- Earliest evidence (a tooth?) of Humans on Earth (in central Israel not Africa)

    400,000 years ago

- Humans began wearing clothing

    170,000 years ago

- Humans migrated out of Africa

    125,000 years ago

- Neanderthals interbreed with Humans

    80,000 to 40.000 years ago

Humans migrated from Asia to North America via Alaska

    15, 000 years ago

- Written Language invented

    10,000 years ago

- Archimedes developed early physics laws and calculus-like methods

    2265 years ago   250 BC

- Isaac Newton revolutionized Physics, Mathematics, and Engineering with his Calculus, Laws of Motion, and Theory of Gravity

    336 years ago   1679

- Charles Darwin's Theory of Evolution

    156 years ago   1859

- James Maxwell's Equations of Electromagnetic Fields

    151 years ago   1864

-Einstein's Special Theory of Relativity

    110 years ago   1905

- Niels Bohr developed one-electron model of atom

    102 years ago   1913

-Einstein's General Theory of Relativity

    100 years ago   1915

- Quantum Mechanics developed

    90 years ago   1925

- Edwin Hubble found the Universe is expanding.

    87 years ago   1928

- Niels Bohr and G. V. Wheeler explained neutron role in fission of Uranium 235

    76 years ago   1939

1. TIMELINE, IMPORTANT CONSTANTS, AND DEFINITIONS (CONTINUED).

- Big Bang Theory of creation of Universe
  65 years ago   1950
- Standard Model of Elementary Particles
  45 years ago   1970
- Inflationary Theory of Creation of the Universe
  45 years ago   1970
- Higgs boson detected to a certainty at CERN
  3 years ago   2012

Supersymmetric particles being searched for at CERN
  2013 and on going

Experimental evidence of inflation of our Universe at origin found in CMBR polarization
  1 year ago   2014
- String Theory/M-Theory of Elementary Particles (Unified Theory of Everything)
  Under development
- Dark Matter and Dark Energy
  Being searched for extensively

The Sun will exhaust its fuel, greatly expand, and burn up the Earth
  7 Billion years from now

The Universe will be completely cold and returned from whence it came
  A few hundred billion years or so from now

#1.2. IMPORTANT CONSTANTS AND DEFINITIONS.

#1.2.1. EINSTEIN'S EQUATION OF THE EQUIVALENCE OF MASS m AND ENERGY E.

$E = mc^2$ Where c is the velocity of light = 2.99 792 458 x 10^8 m/sec^{-1}

This equation shows that it takes only a little tiny bit of mass to make a lot of energy, but it takes lots and lots of energy to make a little tiny bit of mass.

The negative energy it takes to create mass is exactly balanced by the positive energy of the mass that has been created, and vice versa. This principle allows the creation of the Universe (section 17).

#1.2.2. FORCE.
Force F is measured in Newtons. A Force **F** of 1 Newton is defined as the force acting on a 1 kilogram of mass **m** to produce an acceleration **a** of 1 meter per second per second:

$F = ma$

1 Newton =1 kilogram meter per second per second
= 1 kg-meter/sec^2
= 10^5 dynes = 0.2248 pounds = 0.2248 lb

#1.2.3. DISTANCE, WEIGHT, GRAVITY, GRAVITATIONAL FORCE, GRAVITATIONAL FIELD, AND CENTER OF MASS.

1 inch = 0.254 meters = 1 in. = 0.254 m   1 meter = 39.37 inches

1 kilometer = 0.6215 miles   1 mile = 1.609 kilometers

NOTE: Gravity is not a force, but more precisely is the effect of mass on space. This effect is called a force. (See **section 22.**)

The gravitational force g on an object is its weight (w). The weight (force) of an object is its mass (m) times the gravitational force (or other force) of acceleration **g** at the location of the mass.

$$w = mg \quad m = w/g \quad g = 9.81 \text{ m/sec}^2 = 32.102 \text{ ft/sec}^2$$

GRAVITY: Newton's Law of Gravitational Force of attraction F_g between two objects:

$$F_g = (G\, m_1\, m_2)/r^2$$

G is gravitational constant:

$$G = 6.67385 \times 10^{-11} \text{ m}^3 \text{ kg}^{-1} \text{ s}^{-2} = 6.67385 \times 10^{-11} \text{ N m}^2 \text{ kg}^{-2}$$

N is force in Newtons, G is also identified as G_N, where F is the force between the objects (masses) in Newtons (N); m_1 and m_2 are two masses in kilograms; r is the distance between the centers of the two masses in meters. **NOTE CAREFULLY:**
$F_g = (G\, m_1\, m_2)/r^2 = m_2\, a = m_2\, (G\, m_1)/r^2$ **(See section 1.2.2 above.)**

$(G\, m_1)/r^2$ **is termed the acceleration of mass m_2 by mass m_1 . Thus, it doesn't matter if m_2 is a feather or a truck; the acceleration by m_1 is the same. A one ounce weight and a ten ton weight both fall to the Earth at the same acceleration (time) by the Earth!**

GRAVITATIONAL FIELD E_m OF MASS m: $E_m = (G\ m)/r^2$

MASS, SLUGS, POUNDS (LB), AND GRAVITATIONAL FORCE:

Metric Units: **mass** in **kg (kilograms)**
              force in **Newtons (N)** **(kg-m/sec^2)**
English Units: **mass** in **slugs, force** (weight) in **lb (pounds)**.
Metric units have essentially replaced English Units in scientific work.
a. 0.00700 slugs = 0.2248 lb = 0.102 kg = 1.000 N = 1.000 kg-m/sec^2
b. 0.3115 slugs = 1.000 lb = 0.454 kg = 4.448 N = 4.448 kg-m/sec^2
c. 0.6853 slugs = 2.200 lb = 1.000 kg = 9.807 N = 9.807 kg-m/sec^2
d. 1.000 slug = 32.107 lb = 14.59 kg = 143.0 N = 143.0 kg-m/sec^2

Acceleration g due to gravity of Earth (at an elevation of 14 meters)
      $g = G\ m_{Earth}/r^2 = 980.665$ cm /sec^2 = 32.1652 ft/sec^2 where r is distance to center of Earth

CENTER OF MASS: Many times in solving complex problems such as a meteor passing near the Earth and the Moon, calculations can be simplified if the center of mass is of two of the objects is determined and then those two objects can be treated as only one object. This reduces a very complex three body problem to a simpler two body problem.

Finding the center of mass for two objects is like finding the balance point on a teeter-totter. The heavier (more massive) person slides in toward the fulcrum (center) while the lighter person slides to further out from the fulcrum until perfect balance is achieved. The balance point is the center of mass and is calculated as follows:

      $m_1\ r_1 = m_2\ r_2$ $r = r_1 + r_2$
      m_1 is the mass of the first object.
      m_2 is the mass of the second object.
      r is the distance between the centers (of mass) of m_1 and m_2 .
      r_1 is the distance of the first object from the center of mass.
      r_2 is the distance of the second object from the center of mass.

#1.2.4. MASS (IN KILOGRAMS) OF VARIOUS OBJECTS: (See also [Sternheim, p. 44]

Electron	$= 9.109 \times 10^{-31}$	Proton	$= 1.673 \times 10^{-27}$
Oxygen atom	$= 3 \times 10^{-26}$	Ant	$= 10^{-5}$
Humming bird	$= 10^{-2}$	Man (220 pounds)	$= 10^{2}$
Elephant	$= 10^{4}$	Whale	$= 10^{5}$
Ship	$= 10^{8}$	Moon	$= 7 \times 10^{22}$
Earth	$= 5.972 \times 10^{24}$	Sun	$= 1.988 \times 10^{30}$
Milky Way Galaxy	$= 2 \times 10^{41}$	Universe	$= 3 \times 10^{56}$

NOTE: The term "mass" as used in this book usually means the composite of rest mass m_o, relativistic mass m_r, and energy mass m_e. Relativistic mass and energy mass must be considered when extreme circumstances apply such as of velocities approaching the velocity of light and/or large objects such as the Sun, galaxy, or black hole. These situations are discussed in this book where applicable.

#1.2.5 WORK AND ENERGY.

Work (W) on an object is done by a Force (F) moving an object (mass) a distance (s).

Work = Force times distance s.   W = Fs = mas (from **section 1.2.2**)

Work is measured in joules. 1 joule is the work done by a force of 1 Newton moving a 1 kilogram object 1 meter in the direction of the movement. Energy is the ability to-do Work. Work and Energy have the same dimensions.

1 joule = 1 Newton-meter = 1 N-m

$$= 1 \text{ kilogram-meter}^2/\sec^2 = 1 \text{ kg-m}^2/\sec^2$$

1 joule = 10^7 ergs = 1 kg-m^2/sec^2

1 erg = 1 (gm-cm^2/sec^2)

#1.2.6. KINETIC ENERGY.

Kinetic Energy (K) is the ability of an object m to do Work by virtue of its motion (velocity v squared).

$$K = \frac{1}{2} mv^2 \quad \text{Where m = mass, v = velocity}$$

The Kinetic Energy K of a 1 kg mass m moving at 1 m per sec:

$$K = \frac{1}{2} mv^2 = \frac{1}{2} \times 1 \text{ kg} \times (1 \text{ m/sec})^2 = \frac{1}{2} \text{ kg-m}^2/\text{sec}^2$$

The Kinetic Energy of 1 gm moving at 1 cm per sec:

$$K = \frac{1}{2} mv^2 = \frac{1}{2} \times 1 \text{ gm} \times (1 \text{ cm/sec})^2 = \frac{1}{2} \text{ gm-cm}^2/\text{sec}^2$$

#1.2.7. POWER.

Power P is the rate at which work is accomplished - the work W accomplished in a given unit of time s. Power is measured in joule-sec.

$$P = Ws$$

1 joule-sec = 1 watt = 10^7 erg-sec = 0.7376 ft-lb/sec = 0.7376 ft-lb/sec

$$= 1 \text{ (kg-m}^2/\text{sec}^2\text{)-sec} = 1(\text{kg-m}^2/\text{sec}$$

$$= 10^7 \text{ gm-cm}^2/\text{sec} = 10^7 \text{ erg/sec}$$

1 horsepower = 550 ft-lb/sec = 33,000 ft-lb/min = 7.457 x 10^2 watts

1 kilowatt = 1 kw = 1 x 10^3 watts = 1.341 horsepower

#1.2.8. LINEAR AND ANGULAR MOMENTUM.

Linear momentum p is defined as mass m times velocity **v** p = m**v**.

Angular momentum L is defined as the mass m times the instantaneous velocity v of the object times the distance r of the object from the fixed point. (Think of an object m on a string being twirled around your head at a velocity v at a distance away r from your head.)

$$L = mvr$$

#1.2.9. VELOCITY, FREQUENCY, AND WAVELENGTH OF LIGHT.

Velocity of light (c)
  186,000 miles per second
  299,792,458 meters per second = 3×10^8 m/sec

Frequency (v) of photons (**v** is the lower-case of Nu of the Greek alphabet.)
  Frequency v (cycles per second) is expressed in Hertz (Hz) for photons, other electromagnetic energy, and electromagnetic radiation.

Wavelength of photons (λ)
  A wavelength is the distance an electromagnetic wave can travel at the velocity of light in one cycle of the wave.

λ= c/**v** (Velocity of light divided by frequency)

#1.2.10. ENERGY (E) OF PHOTON.

The greater the photon frequency v or the shorter the photon wavelength λ, the greater the energy **E** of a photon:

$E = hv = hc/ \lambda = v(6.626 \times 10^{-27})$ erg-sec
Where h = Planck's constant (See **section 1.2.17**.)

For visible light: $v = 10^{14}$ Hz = 10^{14} /sec
$E = (10^{14}/sec) (6.626 \times 10^{-27})$ erg-sec
$= 6.626 \times 10^{-13}$ erg

#1.2.11. EQUIVALENT MASS OF PHOTON.

A Photon has an equivalent mass as given by Einstein's formula for the equivalence of mass (m) and energy (E): $E = mc^2$ where c is the velocity of light. $m = E/c^2$

For visible light, $m = (6.6 \times 10^{-13}$ erg$) /(3 \times 10^{10}$ meter^2/sec$^2)$
$m = (6.6 \times 10^{-13}$ gm-cm^2/sec$^2)/ (3 \times 10^{12}$ cm^2/sec$^2)$
$m = 2.2 \times 10^{-25}$ gm = 2.2×10^{-28} kg

Although, photon masses are very small, compared to the huge mass of the Sun (2 x 10^{33} gm), the attractive force (F) of gravity is enough to deflect the ordinarily straight path of the photons and bend them toward the sun.

For two masses, m_1 and m_2, a distance (r) apart , the attractive force (F) of gravity is given by: $F= G(m_1 \times m_2)/r^2$ where $G = 6.67 \times 10^{-11}$ N m^2/kg^2

#1.2.12. ELECTROMAGNETIC SPECTRUM OF PHOTONS.

Frequency is measured in Hertz (Hz). 1 Hz – 1 cycle per second (cps)

Household 60 Hz wall power λ = 3100 miles
AM radio 1000 KHz λ = 982 ft (299 meters)
Channel 2 TV (54 MHz) λ = 18.19 ft (5.54 meters)
 Microwave (10^{11} Hz) λ = 2.99 x 10^{-3} meters =2.99 mm
Visible light (10^{14} Hz) λ = 2.99 x 10^{-6} meters
X-ray (10^{18} Hz) λ = 2.99 x 10^{-10} meters
Gamma rays (10^{21} Hz) λ = 2.99 x 10^{-13} meters

Angstrom: $\dot{A} \equiv$ 0.1 nano-meter = 1 x 10^{-10} meters

The Angstrom (named after Swedish astronomer Anders Angstrom) is also used to express electromagnetic wavelengths of spectroscopy.

#1.2.13. COSMOLOGICAL UNITS

Velocity of light   c = 2.99 792 458 x 10^8 m/sec^{-1}
Newtonian Gravitational Constant   G = G_N = 6.6739 x 10^{-11} m^3 kg^{-1} s^{-2}
Light year (ly)   0.946053 x 10^{16} m
Astronomical Unit (au or *A)*   1.495 978 707 00 x 10^{11} m
Parsec (pc)   3.0856776 x 10^{16} m = 3.262 ly = 2.06267 x 10^5 A

First Galaxies form   800 years after Big Bang
Galaxies in Universe   4 x 10^{11}
Stars per galaxy   1 x 10^{12}
Stars in Universe   4 x 10^{23}
New stars now being created by typical Starburst galaxy   2000 per year
New stars now being created by Milky Way galaxy   2 per year

Diameter of Milky Way   6 trillion miles = 6 x 10^{12} miles = 9.6 x 10^{12} km
Mass of Universe   3 x 10^{56} kg
Energy of Universe 10^{50} kg = 9 x 10^{66} joules
Density of Universe 10^{-23} gm/m^3

Age of Universe   13.8 x 10^9 years
Radius of Universe   25 x 10^9 light-years = 25 x 10^9 (0.946053 x 10^{16} m)
                = (2.365 x 10^{26} m
Volume of Universe   = (4/3) (π) (r^3) = (4/3) (3.14159) (6.355314 x 10^{24} m)
                = 1075 x 10^{72} m^3

CMBR temperature   2.725 K

Stars in Milky Way Galaxy   300 billion = 3×10^{11}

Mass of Milky Way Galaxy 2×10^{41} kg

Planets in Milky Way Galaxy   50 billion

Our Solar System, Its Planets, and Pluto formed   4.6 billion years ago

Radius of Sun 6.9551×10^{8} m

Mass of Sun (Msolar) = 1 Solar mass = 1.9885×10^{30} kg

Sun Core Temperature 1.5×10^{7} K

Sun Surface Temperature 6×10^{3} K

Luminosity of Sun (Lsun) 3.828×10^{26} W (watts)

Mean Distance from Sun to Earth   149,597,870,700 m

Mass of Earth 5.9727×10^{24} kg

Radius of Earth 6.378137×10^{6} m

In contrast, here are some units used frequently on Earth:

1 teaspoon (salt)   6.0 grams

1 Cup = 8 fluid ounces = 16 Tablespoons = 48 teaspoons = 237 milliliters

1 C = 8 oz = 16 Tbsp = 48 tsp = 237 ml

$K = 10^{3}$; $M = 10^{6}$; $G = 10^{9}$; $T = 10^{12}$

#1.2.14. AGE AND DIAMETER OF UNIVERSE.

Age of the Universe = 13.8 billion years

The Diameter of the Universe before the BIG bang when the Universe was a tiny sphere crushed by four unified quantum forces =

10^{-10} centimeter - or perhaps as small as 10^{-33} centimeter

Current Diameter of Universe:

40 to 90 billion light years, assuming the Earth is about in the center of the Universe. (See **section 6.4.**)

#1.2.15 ELECTRON VOLT eV, MASS AND CHARGE OF ELECTRON AND PROTON (See also section 12.1.)

1 eV = 1.602 176 565 x 10^{-19} joules (Coulombs)
1 eV/c^2 = 1.782 661 845 x 10^{-36} kg

Electron mass = 0.5109989 Mev/c^2 = 9.109 x10^{-28} grams
Proton mass   = 938.272046 Mev/c^2 = 1.673 x10^{-24} grams

A proton weighs about 1837 times more than an electron.

1 MeV/c^2 = 1.782661845 x 10^{-33} grams

Physicists many times omit the c^2 as understood when specifying mass. (c = the speed of light.)

Charge of electron = -1 eV = -1.602 176 565 x 10^{-19} joules (Coulombs)

Charge of proton = +1 eV = +1.602 176 565 x 10^{-19} joules (Coulombs)

#1.2.16. COSMIC RAYS.

For cosmic rays: Energy E = 10^6 eV to 10^{15} eV.

Energies of cosmic-ray particles are usually between 10^7 eV and 10^{10} eV. Cosmic rays of energies of 10^{15} eV have been detected. These are usually very high velocity protons.

#1.2.17. PLANCK AND BIG BANG SPHERE MEASUREMENTS.

PLANCK'S CONSTANT (h)

$h = 6.26070 \times 10^{-27}$ erg-sec = 6.26070×10^{-34} joule-sec

$= 6.26070 \times 10^{-27}$ (gm-cm^2/sec^2)/sec

$= 6.26070 \times 10^{-27}$ gm-cm^2/sec

$\bar{h} = h/2\pi = = [(6.26070 \times 10^{-27})/ 6.2832]$ erg-sec = 1.054571726 erg-sec

PLANCK MASS: $P_m = (\bar{h}c/G_N)^{1/2} = 1.22093 \times 10^{19}$ GeV/c^2 = 2.17651×10^{-8} kg

Where c is the speed of light.

G_N is the gravitational constant **section 1.2.3.**

Planck length: 1.6162×10^{-33} cm

Planck Time: 3.3×10^{-44} sec

Square Planck: 2.612×10^{-66} cm^2

Cubic Planck: 4.2217×10^{-99} cm^3

Diameter of Planck Sphere (Planck length) = 1.6162×10^{-33} cm

Diameter of Big Bang Sphere (Planck length) = 1.6162×10^{-33} cm

#1.2.18. DIAMETER OF PROTON AND ELECTRON.

Proton is a positively charged Baryon (section 10.3). It can be a free particle. It also is a key part of the nucleus of all atoms:

Diameter of proton = 1.7536×10^{-15} meter = 1.7536 femtometer

Where 1 femtometer = 1×10^{-15} meter = 1×10^{-13} centimeter

Electron is a negatively charged elementary particle (**section 9.2).**

It can be a free particle.

Diameter of free electron = 1.5×10^{-18} meter = 1.5 r x 10^{-3} femtometer

The electron can also act as a wave (**section 12.3**) when it is in orbit around the nucleus of an atom.

#1.2.19. LENGTH, DIAMETER, AND MASS OF VIBRATING QUANTUM STRING OF ENERGY Q_E OF STRING THEORY

Q_E Length = 1.6162×10^{-35} meter = 1.6162×10^{-20} femtometer

$\qquad$ = 1.6162×10^{-33} centimeter = 1 Planck length

Q_E Diameter $\cong 0$

Q_E Energy = 1×10^{-9} eV = 1.6×10^{-28} joules

#1.2.20 de BROGLIE WAVELENGTH λ :

$λ = h/p = h/mv$ where p = momentum = mass x velocity
de Broglie wavelength λ is associated with all mass and matter.
1 gm mass (m) moving at velocity (v) of 1 cm/sec:

$λ = h/mv =(6.626 \times 10^{-27}$ gm-cm^2/sec)/(1 gm x 1 cm/sec)
$λ = 6.626 \times 10^{-27}$ cm ~ 10^{-26} cm

#1.2.21. $π$

$π$ = 3.141 592 653 590

#1.22.22. PRESSURE.

Pressure is another form of energy that can do "Work."
Pressure P is the magnitude of the forces F exerted in all directions of the area A of its
surroundings – pressure is the force F acting on an area A.   **P = F/A**

The basic unit of measurement of Pressure P is the Pascal **Pa. Pressure is Force f**
(Newtons) per unit area A (meters^2)

1 Pascal (Pa) = 1 Newton (N)/ meter^2
1 Pa = 1 N m^{-2}

Normal atmospheric pressure = 1 atmosphere = 1 atm

1 atm =1.013 x 10^5 Pa
        = 1.013 bar
        = 760 mm Hg

The bar (barometric) is used in meteorology. The millimeter (mm) of mercury (Hg) is
used in medicine.

#1.2.23. TEMPERATURE (IN DEGREES).

Centigrade C = 5/9(F -32)   F = Fahrenheit   Kelvin K = C + 273.15

Absolute zero temperature is the lowest possible obtainable temperature. The lowest temperature achieved experimentally is 10^{-9} K. As temperature decreases, elements change phase from Gas, to Liquid, to Solid, to Bose-Einstein Condensate. At temperatures below 2 K, startling effects occur, including superconductivity (zero resistance to electrical current.

Temperature is directly related to ability of Heat Energy to do Work (**section 1.2.5)** and Kinetic Energy (**section 1.2.6).** The hotter (higher temperature) of an object, the more rapidly can heat energy transfer from a hotter to a colder object.

KELVIN (K) TEMPERATURE OF VARIOUS PHENOMENA:

ABSOLUTE zero   0 K   -273.15 C
Temperature of Empty Space 10^{-30} (?) (Beyond our Universe)
Lowest Temperature Achieved 10^{-9} (in a laboratory in our Universe)
Elements Act Like Waves   1
        (Rather than like particles and can join other waves to make big waves)
Bose-Einstein Condensate   2
        Forms with elements all in lowest energy states
Helium Liquefies   4
Dry Ice (CO_2) Freezes   195
Water Freezes (solid)   273.15   0 C
        (Zero degrees Centigrade)
Human Body Temperature   310
Water Boils (gas)   373.15   100 C
Gold Melts   1336
Surface of Sun 6×10^3
Center of Earth 1.6×10^4
Center of Sun 1.5×10^8
Center of a Giant Star 1×10^{10} (20 times bigger than Sun)

38

#1.2.24. RELATIONSHIP OF HEAT TO OTHER FORMS OF ENERGY.

Heat is just another form of Energy. There are many kinds of energy including mechanical, chemical, heat, pressure, mass, nuclear fusion, nuclear fission, and radiant. All types of energy have the ability to do work.

Heat Energy has the ability to do Work (**section 1.2.5**) just as Kinetic Energy (**section 1.2.6**) has the ability to do Work. In the early 1800's, a special set of units were developed for Heat Energy instead of using the ones for Work and Kinetic Energy as it was not understood that Heat is just another form of energy. At that time, the gram-calorie (cal) was defined as:

1 gram-calorie = heat required to raise the temperature of 1 gram of water from 14.5 to 15.5 degrees Centigrade.

In 1842, Julius Robert Mayer suggested that all energies were equivalent and one could be transformed into another. He also suggested that energy cannot be created nor destroyed. Soon after James Prescott Joule and (independently) Hermann von Helmholtz determined that energy is conserved in any interaction – the energy entering into any interaction may be changed from one type to another, but the total energy is always the same at the finish. This continues to be a stalwart principle of all physics.

#1.2.25. ENERGY CONVERSION LIST. (SEE ALSO SECTIONS 1.2.5 AND 1.2.6.)

1 joule = 10^7 ergs = 0.2390 calories = 0.7376 ft-lb
1 calorie = 4.184 joules
1 joule = 6.24×10^{18} electron volts (eV)
1 electron volt = 1.602×10^{-19} joules
1 kilowatt-hour (kwh) = 3.6×10^6 joules
1 British thermal unit (BTU) = 1.054×10^3 joules
1 ft-lb = 1.356 joules
        At K (Kelvin Temperature) = 0   C (Centigrade) = -273.15

#2. BASIC PHYSICS OF THE EARTH AND UNIVERSE.

#2.1. MASS AND ENERGY.

The mass of an object is defined as its inertia, the amount of its resistance to acceleration (change in velocity). An object (mass) at rest tends to stay at rest; an object in motion tends to stay in motion. The Higgs fields (described in detail in **section 13.7**) interact only with objects and particles that are perceived by the Higgs fields to carry mass. Further, the Higgs fields interact only with particles that are changing velocity (accelerating or decelerating). This is analogous to electrically charged particles that radiate electromagnetic energy when accelerating or decelerating. The action of the Higgs fields to resist a change in the velocity of a particle that has mass (energy) gives the particle mass. Without the Higgs fields, particles would not have mass. Particles that are accelerated gain energy and thus gain mass. Particles that are decelerated lose energy and thus lose mass.

Mass and energy are the same thing. (See **section 1.2.1**). Mass is the energy of a particle. Rest mass is the rest energy of a particle. At the fundamental level, it is energy that is conserved in particle interactions. A particles decay or are created, energy is always conserved. As the energy of a particle increases, the entire increase will be physically embodied as an increase of mass, and vice-versa. A measurement of the total mass of a particle in effect is a measurement of the total energy of a particle.

Mass in nature is quantized. It does not exist as a continuum, but rather as a composite of very small discrete values (quanta). As such, a mass and energy cannot continuously be subdivided below these fundamental quanta of energy.

Large masses and small masses are accelerated the same amount by a gravitational field. A heavy and light mass, if dropped together, both free fall at the same rate. This is understandable if a mass is considered to be a composite of energy quanta - each energy quanta is accelerated the same amount by the gravitational field whether in a large mass or small mass.

#2.2. MASS AND ENERGY ARE QUANTA OF SAME THING.

Whew! The previous paragraph got right to the crux of things. But not to worry! There is a whole book ahead of you to clarify these issues starting with this **section**. Let's get started with a question:

What, you may ask? Mass and energy are the same thing? Some readers may have heard about Einstein's famous equation of the mathematical equivalence of mass and energy: $E = mc^2$

This equation is also called Einstein's equation. However, the term "Einstein's Equation" more properly pertains to his equation of General Relativity that extended his Special Relativity (**section 21**) pertaining to motion at constant velocities approaching the speed of light to accelerated motion, gravity, mass, energy, spacetime, and the Universe (**section 22**).

There will be a lot more about this later on in this book. But what's a complex concept like the equivalence of mass and energy doing at the beginning of this book?

It's here at the beginning because proving in simple descriptive terms that mass and energy are the same thing right now, without any mathematics, establishes concepts that are vital to understanding the material in later **sections** about particle physics and the Universe we live in. That's what this book is about.

To begin, select any mass. You know what a mass is – it's an object. Select a big or little object - a piece of paper, a pencil, a car, a planet. Here's a magic knife -cut the object in two. Throw half away. Do it again and again.

No matter how big an object you started with, after a few thousand on cuts or so, you will soon get down to a molecule or atom. Don't stop, keep cutting in half.

2.2. MASS AND ENERGY ARE QUANTA OF SAME THING (CONTINUED).

Now you are down to cutting an atom into two parts: electrons, quarks, and gluons – there is lots more about these and other elementary particles in **Part II** of this book. At this level, every elementary particle such as an electron is exactly alike and interchangeable with every other electron, and similarly for quarks, and gluons. Exactly, exactly alike, not just about alike, exactly alike. Keep cutting in two, you're almost done! Unfortunately, you are beyond the reach of current technology to experimentally confirm the results of the next few cuts. But let's go ahead anyway.

The next few cuts hypothetically divide the elementary particles into vibrating packets or strings of pure energy. Quickly we reach a limit where the packets of energy can't be divided in two again. Unlike dividing a number in two which we can do an infinite number of times (3/2, 3/4, 3/8, ….), we have reached the limit of how many times we can divide an object in two. These vibrating packets of pure energy can only be subdivided so many times into a fundament vibrating packet or string, but not any further.   Let's call this packet or string a "quantum" of energy.

According to "String Theories", this final fundamental quantum of energy can make up other vibrating string energy packets either by vibrating in different modes or combining with other fundamental quanta to form more complex energy packets.   But no matter how hard we try, we cannot divide this fundamental quantum of energy any further.

By our thought process of continuously dividing objects into two parts, we have arrived at a realm that is beyond the reach of current experimental apparatus for confirmation. So far the results are justified for now only through our conjecture. However, we will later justify our conclusions by extrapolating our results to experiments that we can perform.

You may still not be satisfied and claim we divided mass into subparts, but not energy, at least to your satisfaction. Okay, let's go grab some energy, some heat energy for instance and divide it in two parts successively. It's hard to get our hands around some heat energy without getting burned, but there is a way described in detail in **section 4.**

The conclusion reached there is the same: there is a fundamental quantity of heat that cannot be divided further. It's called a heat photon. It's an electromagnetic photon just as light, radio waves, x-rays, and so forth all are electromagnetic photons. The photon is the quanta of electromagnetic energy. The graviton is the quanta of gravitational (mass) energy. The graviton has not yet been detected experimentally.

Mass and energy at their lowest level are quanta of energy that cannot be further subdivided. Can a photon be divided further? Maybe, photons can be divided further into quantum vibrating strings of energy as discussed in earlier in this **section.**

So when you look at any object or its interactions with other objects such as a bat hitting a baseball or a kettle being heated on a stove, remember that at the lowest fundamental level, huge amounts of exactly identical quanta of energy are interacting with each other.

Here are some very important fundamentals:
a. Mass and energy are equivalent.
b. Energy and mass quanta are the same thing. What we think of as a mass quantum is the same as an energy quantum and what we think of as an energy quantum is the same as a mass quantum.
b. Energy/mass quanta can combine and interact with other mass/energy quanta to produce all the elementary particles, atoms, molecules, planets, stars, galaxies, and energy in the Universe.
c. Energy and mass are quantized and a quantum of energy cannot be subdivided.
e. Energy and mass are conserved. If 100 quanta of mass and energy are involved at the beginning of an interaction, there will be 100 quanta after the interaction. Some mass quanta may have changed to energy quanta and vice versa.

Finally, the reader may be wondering where all the mass and energy quanta in the Universe came from. That's the story for the rest of this book. Suffice it to say at this point, "They were borrowed, but what was borrowed must be paid back."

Experimental evidence for the quantum theory of matter and a description of the origin of this theory is given **section 4.2.**

#2.3. FORCE, MASS, GRAVITY, AND WEIGHT.

#2.3.1. FORCE.

Force is a push or pull on an object. An object set into motion by a force will be accelerated in the same direction as the force. The amount of acceleration is proportional to the magnitude of the force. Force is a vector that has both magnitude and direction. (Vectors are usually shown in **bold face** type.)

The net Force **F** is related to the mass m of an object and its acceleration **a** by Newton's (second) law: **F** = m**a**

Force F is measured in Newtons. A Force f of 1 **Newton** is defined as the force acting on a **1 kilogram** of mass and accelerating it to **1 meter per second per second**

   1 Newton =1 kilogram meter per second per second = 1kg-meter/sec^2
      = 1kg-m-sec $^{-2}$

1 meter per second per second means that the kilogram being accelerated travels 1 meter in the first second, 2 more meters in the second, 3 more meters in the third second, and so forth. The 1 kg mass goes faster, and faster, and faster, until stopped by force acting in the opposite direction or a barrier of some kind is encountered.

Contact forces are exerted when the objects involved are in contact. For example, the force you exert when pushing a stalled car. Gravitational, magnetic, and electrical forces can be exerted by objects not in contact. For example, the force of gravity keeps the Earth in orbit around the sun.

If you continue to push a car with a constant amount of force, it will continue to accelerate and you will not be able to keep up with it. In the actual situation, the "dissipative" forces of friction will prevent the acceleration from continuing even with a constant force being applied. See **section 2.5.2** for more information about dissipative and conservative forces.

The metric unit of force is the Newton (N). The American unit of force is the pound (lb). See below.

1 Newton = 10^5 dynes = 0.2248 pounds = 1 kilogram-meter/sec^2

#2.3.2. WEIGHT.

The gravitational force g on an object is its weight (w). The weight (force) of an object is its mass (m) times the gravitational force (or other force) of acceleration **g** at the location of the mass.

$w = mg \quad m = w/g$

The mass of an object is invariant while its weight varies with the strength of gravity where the object is located. An object weighs far more on Earth than on the moon, but its mass is the same at both locations. Ordinarily, objects are weighed at the surface of the Earth where:

$g = 9.81$ m/sec^2 = 32.107 ft/sec^2 (sec^2 means sec per sec.)

The metric unit of mass is the gram (g) and the kilogram (kg). The American unit of mass is the slug which is rarely used. In America, the mass of an object is generally referred to by its weight in pounds at the gravity at surface of the Earth (**g** = 32.107 ft/sec^2).

#2.3.3. GRAVITY AND GRAVITATIONAL WAVES.

.

Gravitational Force of attraction F_g between two objects m_1 and m_2 is:

$$F_g = (G_N m_1 m_2)/r^2$$

Where F is the force between the objects in Newtons (N), m_1 and m_2 are two masses in kilograms, r is the distance between the centers of the two masses in meters.

G_N is the gravitational constant:

$$G = 6.67 \times 10^{-11} \text{ N m}^2 \text{ kg}^{-2}$$

 The concept of a gravitational field (and an electric field) is useful in many instances later in this book.

The gravitational field E_m of a mass m is defined as:

Gravitational Field $E_m = (G_N m)/r^2$

Compare this equation with the equation above for the gravitational force of attraction between two masses m_1 and m_2 . If the gravitation "field" of a mass is known at any point r, consequently, the gravitational effect of the mass on any other mass at any distance r away from it is also known. The concept of "fields" has played a significant part in developing the theories of physics in particular particle physics and cosmology which will be discussed later in this book.

Although the above equation for gravitational attraction **F_g** (attributed to Isaac Newton), is accurate for most uses, it considers only what is called rest mass. As Einstein's mass-energy equation (**section1.2.1**) shows, to be accurate, mass must include all kinds of mass due to energy (rest, potential, kinetic, chemical, heat, electromagnetic, etc.). This is the great contribution of Einstein's General Theory of Relativity.

Anytime there is a change in the energy of a mass, its mass and gravitational field will vary accordingly, producing **gravitational waves**. An explosion of a giant supernova star, for example, produces gravitational waves that affect you. Or heating a kettle - as the kettle takes on more energy as it absorbs heat and has more mass. However, as it takes so much energy to produce a minuscule amount of energy mass (**section 1.2.1**), you will not ordinarily ever detect the change in gravity without very sensitive instruments.

Gravitational waves are thought to consist of gravitons. Individual gravitons have not been detected experimentally. Evidence of gravitation waves has been detected in the Cosmic Microwave Background Radiation (CMBR). See **section 6.7.**

NOTE, VERY IMPORTANT: Gravity is an attractive (positive) force or energy. Gravitational Energy is negative when it takes energy to separate two objects pulled or held together by gravity. It takes negative energy for you to jump up from positive gravity energy holding you down! Gravitational energy is positive when two objects are attracted.

It takes negative gravitational energy to raise a pendulum to its high point. It takes positive gravitational energy to pull the pendulum down to its low point. The energy at the high point of a pendulum is potential energy – it took negative energy to move the pendulum there that is now stored as positive potential gravitational energy.

Release the pendulum, the stored positive potential gravitational energy is now released by the downward kinetic energy of motion of the pendulum (section 2.5.1).

As the pendulum swings back upward by the kinetic energy of its motion, its kinetic energy is converted back to potential energy at the top of the pendulum's swing.

Positive gravitational (potential) energy pulled the pendulum down.

Negative gravitational (kinetic) energy pulled the pendulum up.

The energy involved is neither created nor destroyed – it just changes from positive potential energy to negative kinetic energy and back again and again as the pendulum swings back and forth.

#2.3.4. UNITS OF MASS AND FORCE.

English Units: **MASS** in **slugs** ; **FORCE** (Weight) in **lb (pounds)**

Metric units have essentially replaced English Units in scientific work.

Metric Units: **MASS** in **kg (kilograms)**; **Force** in **Newtons (N)**: **(kg-m/sec^2)**

a. 0.00700 slugs = 0.2248 lb = 0.102 kg = 1.000 N = 1.000 kg-m/sec^2
b. 0.3115 slugs = 1.000 lb = 0.454 kg = 4.448 N = 4.448 kg-m/sec^2
c. 0.6853 slugs = 2.200 lb = 1.000 kg = 9.807 N = 9.81 kg-m/sec^2
d. 1.000 slug = 32.107 lb = 14.59 kg = 143.0 N = 143.0 kg-m/sec^2

Acceleration g due to gravity(at an elevation of 14 meters)

$$g= 980.665 \text{ cm /sec}^2 = 32.1652 \text{ ft/sec}^2$$

For mass (in kilograms) of various objects, see **section 1.2.4.**

#2.3.5. NEWTON'S LAWS OF MOTION.

Sir Isaac Newton (1642-1727) was the first to develop comprehensive laws of motion which described quantitatively the effects of force including gravity. These laws continue to provide extremely accurate results except at relative velocities that approach the speed of light and at mass, energy, gravity, and acceleration conditions that are extreme. The ramifications of these three laws will arise in many places throughout this book.

NEWTON'S FIRST LAW.

Every object continues to be at rest or continues to move in a straight line unless acted upon by a force. This property is called inertia. Mass is the measure of Inertia of an object. An object that is difficult to start in motion has large mass and high inertia. An object that is difficult to start in motion is likewise difficult to stop.

Later **section**s more comprehensively define different kinds of mass under various conditions:

(1) The original mass of a particle or object when it was created or is at rest (rest mass m_o) - **section 13.7**.

(2) Increasing relative velocity increases mass of an object as velocity of an object increases to approach the velocity of light (relativistic mass m_r) - **section 21**.

(3) Gaining energy by heat, pressure, or other forms of energy increases mass of an object (energy mass$_e$). Subjecting a mass to acceleration of any kind including by massive objects (moons, planets, stars, black holes, and galaxies) increases mass of an object. In this case, objects in free-fall no longer will longer follow "straight lines" but will follow curved geodesic paths that are calculated by the mathematics of Einstein's General Relativity - **section 22**.

NEWTON'S SECOND LAW.

When an object is subjected to a net force, the object will accelerate in the direction of the net force. Section **1.2.2** defines force and **section 2.3.5** gives Newton's Second Law:

$$F = ma$$

A Force **F** of 1 Newton is defined as the force acting on a 1 kilogram of mass **m** to produce an acceleration **a** of 1 meter per second per second in the direction of the net force.

1 Newton = 1 kilogram meter per second per second = 1 kg-meter/sec^2
       = 10^5 dynes = 0.2248 pounds = 0.2248 lb

NEWTON'S THIRD LAW.

If an object exerts a force on a second object, the second object exerts an equal but opposite force on the first object.

For example, if you push away from the wall of a swimming pool with your legs, you move away, but the wall remains motionless - but nevertheless the wall exerted an equal but opposite force on you. If you push away on another swimmer (or a small boat), the other swimmer or boat will recoil in the opposite direction.

#2.4. WORK.

Work (**W**) on an object is done by a Force (F) moving an object (mass) a distance (**s**) at an acceleration **a**.

Work = Force **F** times distance **s**.   W = **Fs** = m**as** (from **section2.3**)

Work is measured in joules. 1 joule is the work done by a force of 1 Newton moving a 1 kilogram object 1 meter in the direction of the movement.

  1 joule = 1 Newton-meter = 1 N-m

       $= 1$ kilogram-meter^2/sec$^2 = 1$ kg-m^2/sec^2

  1 joule $= 10^7$ ergs = 1 kg-m^2/sec^2

#2.5. ENERGY .

Energy is the ability to do Work. The two types of energy are kinetic and potential. Forces involved with energy are conservative and dissipative. Energy and Work have the same units (kg-m^2/sec^2). **Energy** can be positive or negative (**section 2.3.3**). **Pressure** can also be positive or negative (**section 2.10**).

#2.5.1. KINETIC ENERGY.

Kinetic Energy (K) is the ability of an object m to do Work by virtue of its motion (velocity v squared).

$$K = \tfrac{1}{2} mv^2 \qquad \text{Where m = mass, v = velocity}$$

The Kinetic Energy of a 1 kg mass m moving at 1 m per sec:

$$K = \tfrac{1}{2} mv^2 = \tfrac{1}{2} \times 1 \text{ kg} \times (1 \text{ m/sec})^2 = \tfrac{1}{2} \text{ kg-m}^2/\text{sec}^2$$

The Kinetic Energy of 1 gm moving at 1 cm per sec:

$$K = \tfrac{1}{2} mv^2 = \tfrac{1}{2} \times 1 \text{ gm} \times (1 \text{ cm/sec})^2 = \tfrac{1}{2} \text{ gm-cm}^2/\text{sec}^2$$

The final Kinetic Energy K_f of an object m is its initial kinetic energy K_i due to its initial motion plus the Work W done on it (mas) by any Forces F (F = ma) acting on it as it moves a distance s.

$$K_f = K_i + W = \tfrac{1}{2} m(v_i)^2 + \mathbf{mas}$$

The work done on an object is equal to the change in its kinetic energy. If no work is done, there is no change in kinetic energy – energy is conserved.

The final kinetic energy of an object is equal to its initial kinetic energy plus the work on the object as it moves a distance s.

If the initial kinetic energy K_i decreases to zero, all its energy is converted to work. Energy is conserved!!!

See also section 2.3.3.

Note how natural it seems to conclude mass M and energy E are equivalent from the above equation for Kinetic Energy. $K = E = \tfrac{1}{2} mv^2 \qquad M = 2 E v^2$

Einstein was the first to do so. See **section 1.2.1.**

#2.5.2. POTENTIAL ENERGY, CONSERVATIVE FORCES, AND DISSIPATIVE FORCES.

An important kind of work W (**section 2.4**) is the work done by a Conservative force. If work W is done by a conservative force, then the work done will be exactly returned – perpetual motion!

For example, work W (W = - mgh) is done to move a mass m of a heavy pendulum m against the force mg of accelerative gravity **g** to its highest point h.

The acceleration **g** of the force of gravity F_g on a mass m is:

$$F_g = mg$$

Note that the path the pendulum travels is a curved path traveling both horizontally as well as vertically to a height h. Moving against gravity horizontally doesn't do any work against gravity as the horizontal motion is perpendicular to gravitational attraction.

For conservative forces, the paths taken don't matter – only the distance traveled parallel to the accelerative force.

 At the high point of the pendulum, **potential energy** has been stored in the pendulum as the force of gravity tries to accelerate it downward.   When the pendulum is released and moves downwards, the work done against gravity to raise the pendulum now is returned as gravity causes the pendulum to accelerate and swing down to its original rest position.

At its lowest point, all the potential energy of the work done to raise the pendulum is now returned as kinetic energy. From the formulae at the end of the previous **section**:

$$K_f = K_i + W = \tfrac{1}{2}\, m(v_i)^2 + mas$$

Where K_f and K_i are the initial and finial kinetic energies.

$$K_f = 0 + W = - mgh = - \tfrac{1}{2}\, m(v_f)^2$$

Therefore: $\tfrac{1}{2}\, m(v_f)^2 = - mgh$
$$(v_f)^2 = 2gh$$
$$v_f = (2gh)^{1/2}$$
$$\tfrac{1}{2}\, m(v_f)^2 = - mgh$$

The accelerated motion of the pendulum due to gravity has now been converted entirely to kinetic energy (- mgh) at its lowest point. The kinetic energy causes the pendulum to swing up on the other side and stop momentarily at the top of its swing. Kinetic energy has again been converted to potential energy at this point. Gravity now takes over again and causes the pendulum to move downward again. The process is repeated and the pendulum swings up and down.

Is this perpetual motion? Almost yes, depending how successfully dissipative forces of friction were removed, which can never be done for a swinging pendulum. However, there are many instances in physics where dissipative forces are essentially eliminated or can be neglected, if not entirely eliminated.

Unlike conservative forces which are independent of the path of an object and its twists and turns, dissipative forces are dependent on path.

A conservative force may take different paths to get to a destination, but on some paths work is done and on others work is returned. In net, the only issue is the net path parallel to the applied force. If an object is moved back and forth, the net work is zero and work is done in one direction and returned on the return path. Work may not always be returned, but is always available to be returned.

It takes work to hike to the top of a ski run. The work is returned as you ski down. You have a great amount of kinetic energy at the bottom, but unfortunately no way to utilize it. That's not the fault of the physics involved! Rube Goldberg always found a way to harvest unused work of kinetic energy in his machines.

Dissipative forces such as friction do not return work to a system, they extract it. There the work depends on the path and all paths are dissipative. Friction opposes motion. Usually the dissipative force of friction converts the energy of a system into unrecoverable heat energy.

#2.5.3. PRINCIPLE OF CONSERVATION OF ENERGY DETERMINES VELOCITY REQUIRED TO ESCAPE FROM EARTH.

The principle of conservation of total energy is used in explanations in many places in this book. It is a key principle of physics and chemistry. Total energy E is the sum of kinetic K and potential U energies. Total energy is always conserved less any energy dissipated by friction, heat, or otherwise (**section 2.5.2**).

Drop any object. Watch as the potential energy of the object due to gravity is converted to kinetic energy of motion as the object accelerates and falls toward the floor or the ground.

Newton's law of Gravitational Force F_g of attraction between an object m_o a distance r from the Earth and the Earth M_E (**section 1.2.3**) is:

$$F_g = (G\, m_1\, m_2)/r^2 = (G\, m_o\, M_E)/R_E^2$$

G is a gravitational constant:
$$G = 6.67385 \times 10^{-11}\ m^3\ kg^{-1}\ s^{-2} = 6.67385 \times 10^{-11}\ N\ m^2\ kg^{-2}$$

The gravitation force **g** on an object m_o at the surface of the Earth is its weight w_o in the gravitational field **g** on the surface of the Earth (**section 1.2.2**). It is also the potential energy U of the object.

The gravitational potential energy U of an object M acting on a mass m a distance away R is :
$$U = (G\, m\, M_E)/R$$

Note: As explained below, an object m_o will escape from the surface of the Earth if it has enough kinetic energy K to overcome the potential energy U of the gravity of the Earth's mass M_E at the Earth's surface R_E. The potential energy U of the Earth's gravity at the Earth's surface R_E is:

$$U = w_o = m_o g = m_o\ (9.81\ m/sec^2) = (G\, m_o\, M_E)/R_E$$

Ordinarily as two objects fall toward each other, their velocity increases toward infinity as their distance apart decreases. However in the case of an object escaping the Earth, the object and the Earth move further apart and the object's velocity and kinetic energy eventually reduce to zero

To escape from the confines of the gravity of the Earth and conserve total Energy, the kinetic energy K of an object m_o escaping at a velocity v_o from the gravity of the Earth's mass M_E at the Earth's surface R_E must be equal to or greater than the Earth's potential energy U.

$$K = \tfrac{1}{2}\, m_o\, v_o^2 \geq U \geq (\mathbf{G}\, m_o\, M_E)/R_E$$

$$v_o^2 \geq 2(\mathbf{G}\, M_E)/R_E$$

$$v_o \geq [\,(2\mathbf{G}\, M_E)\,/R_E\,)]^{1/2} = (2g\, R_E\,)^{1/2}$$

This is the velocity that provides enough kinetic energy to overcome the potential energy of the Earth's gravitational field. Any object with this velocity will escape from the Earth if its path is directed away from the Earth.

To escape from the Earth's gravity, an object at the Earth's surface requires a velocity of more than 40,000 kilometers per second (25, 000 miles per hour). This is regardless of the mass of the object – escape velocity is independent of the mass of the object. Nevertheless, the kinetic energy required for an object to attain escape velocity is dependent on the mass of an object and for a heavy object requires enormous energy. To determine the velocity required to escape from any other planet or object, substitute its mass for the mass of the Earth M_E.

Light, which moves at the velocity of light c always has enough velocity to escape from any planet, star, or massive object except a black hole. Light is slowed by a gravitational field, and in the case of a black hole, this slowing is enough to slow down light and prevent it from attaining escape velocity – hence, the term, "black hole."

#2.6. POWER.

Power P is the rate at which work is accomplished (the work accomplished in a given unit of time). Power is measured in watts or horsepower.

1 joule-sec = 1 watt = 10^7 erg-sec

$$= 1 \, (kg\text{-}m^2/sec^2 \,)/sec = 1(kg\text{-}m^2/sec^2$$

$$= 10^7 \, gm\text{-}cm^2/sec = 10^7 \, erg/sec$$

1 erg = 1 $(gm\text{-}cm^2/sec^2)/sec$

1 horsepower = 550 ft-lb/sec = 33,000 ft-lb/min

#2.7. LINEAR MOMENTUM.

So far, the topics of mass m, mass and velocity squared (kinetic energy mv^2), mass and force of acceleration **a** have been covered. In these topics, energy was conserved and by following the energy, numerous problems can be solved and results of experiments predicted.

Nevertheless, simple mass and energy mv have been neglected until now. Mass m times velocity **v** is called momentum **p.**

   p = m**v**

Both **p** and **v** are vectors and have magnitude and direction. Momentum is very important to physics (and especially in particle physics) because it is conserved in many instances. Like energy, the momentum at the beginning of an interaction is exactly the same after the interaction (at least in many cases). The good news is that if the momentum is known at the beginning of an interaction, it is also known at the end of the interaction, and in many cases, no other information is necessary to solve a problem.

Cases of interaction where momentum is conserved are called elastic collisions such as between two billiard balls. Cases where momentum is not conserve are called inelastic collisions such as between two cars where much energy is expended in damage to both vehicles. Nevertheless, knowledge o f the momentum of both vehicles provides information with which to compute numerous aspects of the collision based on investigation of the accident.

The only cases of pure elastic interactions are between molecules, atoms and elementary particles providing the energies involve are not too high.

#2.8. ANGULAR MOTION AND ANGULAR MOMENTUM.
Angular motion is the rotation of a mass (object) around a fixed point - for example, whirling a mass on a rope around your head, or an ice skater or ballet dancer whirling around.

Angular momentum L is defined as the mass m of the object times the instantaneous velocity v of the object times the distance r of the object from the fixed point.

$$L = mvr$$

It is also defined in terms of moment of inertia I of the rotating body times the angular velocity ω.

$$L = I\,\omega.$$

A discussion of moment of inertia (I) and angular velocity (ω) is contained in the general physics references in **the Bibliography.**

Angular momentum is important as it is conserved in most interactions.

#2.9. TRANSFER OF HEAT ENERGY.

Heat energy transfer occurs from regions and objects of higher temperature to regions and objects of lower temperature. The disparate temperatures of two regions and objects will eventually approach the same temperature. Heat energy can be transferred by conduction between objects in contact, by convection motion of the objects (liquids and/or gasses) intermingling, and by radiation of photons from one object to another. As explained below, at the atom level, these three methods are not very different. In fact, in many cases, two or all three methods of transfer of heat energy may take place at once.

At the atom level, temperature is a measure of the excitation of the electrons in various orbits around the protons (and neutrons) in the nucleus of the atom. At a constant temperature, electrons are continually jumping between higher and lower orbits. The rate at which this occurs is dependent on the constant temperature of the atom (or molecule). The higher the atom constant temperature, the more "excited" the atom is and the more often the electrons jump back and forth between a higher and lower orbit. This does not change the constant temperature of the atom as the average orbital energy location of all the electrons in an atom remains exactly the same.

As the electrons drop to a lower energy orbit, photons of the commensurate energy are emitted. In turn, the emitted photons are absorbed by another electron, causing it to move to a higher orbit. At constant temperature (absorption temperature), no net photons are radiated from the atoms or absorbed by the atoms.

As an atom is heated, the temperature of the atom increases and the electrons absorb photons from the heat and jump to higher energy orbits commensurate with the higher temperature with a net absorbing of energy by the atom.   Then as the atom reaches the new higher constant temperature, the electrons resume jumping back and forth between a higher or lower orbit but at an increased rate – the atom is more "excited." As the temperature continues to increase further, the negative electrons would eventually have enough energy to entirely escape from the attraction of the atom's positively charged proton nucleus - then the atom would turn into positively charged ion. What happens at even higher temperatures is described in **section 6.2.**

The specific frequencies of the photons absorbed by the electrons of an atom (or molecule) at constant temperature give rise to the black "absorption lines in the spectrum **(section 4)** of an atom which can specifically identify the material being heated.

Conversely, as the temperature decreases, the electrons drop to lower orbits commensurate with the lower temperature and heat photons are radiated from the atom also shown by the "spectrum" of the atom **(section 4).** At this lower temperature the rate at which the electrons jump between higher or lower orbits decreases. As absolute zero is approached, motion of electrons decreases and finally stops if absolute zero temperature could be achieved. However, some materials that are specially fabricated semiconductors (with conductivity between that of an insulator and a good conductor) exhibit "superconductive" characteristics as absolute zero is approached. The jitter of an excited atom is caused by the momentum of electrons moving between higher and lower orbits in an atom due to photons being absorbed by and radiated from an atom. The amount of excitation of the atom is determined by the momentum of the photons involved. Jitter in one instance would be for photons being absorbed or emitted.

#2.9.1. TRANSFER OF HEAT ENERGY BY CONDUCTION.

If two objects of different temperatures T_1 and T_2 are connected by a rod of area A and length l with a thermal conductivity k, the temperature difference T_2 - T_1 will gradually diminish to zero. The time it takes for the heat energy of the hotter object to flow to the cooler object is dependent on the area, length, and (perhaps most important) the thermal conductivity k of the rod. Some example conductivities are silver 420, copper 400, steel 79, glass 0.8, wood 0.08, air 0.024, down 0.019. In general, materials that are good insulators or conductors of electricity are also good insulators or conductors of heat.

It is not necessary for a rod to be between the two objects, so long as they are in contact with each other.

#2.9.2. TRANSFER OF HEAT ENERGY BY CONVECTION.

Convection is the transfer of heat energy by actual motion of hot liquids and/or gasses. If any gas or liquid material is heated it will take on energy and expand (and also have more mass but not very much more). The motion of convection will "stir" the disparate temperature gasses or liquids together and equalize their temperatures. A solid rod will get longer and weight more (**section 4.1**). A liquid or gas will expand if heated and become less dense. If not allowed to expand by the container, the pressure of the liquid or gas in the container will increase, and if heated enough, will burst the container (**section 2.10**).

If a liquid in a pot is heated, the hotter liquid at the bottom of the pot will expand becoming lighter than the cooler liquid at the top of the pot. The hotter liquid will rise and replace the colder, heaver liquid above it. When the warmer liquid reaches the top of the container, it cools, contracts, becomes denser, and sinks toward the bottom of the pot. This process continues with hotter fluid moving to the top and cooler fluid moving to the bottom to equalize the disparate temperatures.

 Heat energy transferred by the differences in liquid density as described for a heated pot is called natural or free convection. Heat energy transferred by a pump, blower, or stirring is called forced convection.

 If the liquid at the top of the pot is heated, it will not sink by free convection and the liquid at the bottom will be heated by conduction as described in the previous **section**.

There are many examples of heat transfer by convection.
- Hot air from a furnace rises to the upper level.
- Hot air balloons heat the air in the balloon so it is light enough to carry passengers aloft.

- A very interesting example of convection is the cooling of water. As water cools, it becomes denser and as expected cooler water sinks to the bottom of a lake, the ocean, or an ice cube tray. Warmer water rises to the top. Water at the top of a swimming pool, a lake, the ocean, and an ice cube tray is warmer than water at the bottom. Like most substances, **water increases in volume as water temperature increases** and less dense hotter water floats on top of cooler water.

However, water is unusual. When water changes "state" (from liquid to solid), it no longer decreases in volume as temperature decreases. **Water increases in volume as temperature of water decreases from 4 C (degrees Centigrade) to 0 C and colder water floats on top of warmer water in this temperature range.** This is why swimming pools, lakes, the ocean, and ice cube trays freeze at the top first – otherwise, all the ice would be at the bottom perhaps never melt or melt very slowly.

#2.9.3. TRANSFER OF HEAT ENERGY BY RADIATION.
Transfer of heat energy by radiation (**section 1.2.12**) refers to the emission and absorption of heat energy by photons. Transfer of heat energy by radiation is discussed in detail in many of the later **sections** of this book.

#2.9.4. CHANGE OF STATE.

All substances change "state" from solid, to liquid, to gas as they are heated by conduction, convection, or radiation:. Water changes state from solid (ice), to liquid (fluid) state, to gas (water vapor) state. Then as the substance cools, the process is reversed. Gaseous water vapor cools to liquid drops perhaps of dew. Liquid water freezes to solid ice.

In some substances, the liquid state is fleeting: solid wood or coal burns to gaseous smoke without apparent transition to liquid form. In many instances we may encounter some substances only as solid, liquid or gas.

Water expands as it is heated (**section 2.9.2**). It takes 1 calorie of heat added to 1 gram of water to increase its temperature by 1 degree C (Centigrade). It takes 540 calories to heat 1 gram of water at 100 deg C and change its state from liquid into gas (steam). This great heat capacity of calories of steam is why steam burns are much more dangerous than boiling water burns.

Below 4 deg C, water expands (becomes less dense) as it freezes, otherwise ice would sink to the bottom of a lake or ocean and perhaps never melt. Eighty (80) calories must be removed from each 1 gram of water at 0 deg C to change it into 1 gram of ice at 0 deg C. Conversely, 80 calories must be added to each gram of ice at 0 deg C to change it into 1 gram of water at 0 deg C.

#2.10. PRESSURE.

Pressure (P) is very important in this book. Pressure is another form of energy that can do "Work." Since pressure is a form of energy, it is also has "gravity," just as mass has gravity. It is of particular importance in the creation of our Universe during its inflationary expansion phase.

"Positive" Pressure (Positive Gravity) and "Negative" Pressure (Negative Gravity) are important to the birth of our Universe (section 6.2).

Pressure P is the magnitude of the forces F exerted in all directions of the area A of its surroundings – pressure is the force F acting on an area A: $\qquad P = F/A$

#2.10.1. UNITS OF MEASUREMENT OF PRESSURE.

The basic unit of measurement of Pressure P is the Pascal **Pa.** **Pressure is Force f (Newtons) per unit area A (meters^2)**

$$1 \text{ Pascal (Pa)} = 1 \text{ Newton (N)}/ \text{meter}^2$$
$$1 \text{ Pa} = 1 \text{ N m}^{-2}$$

Normal atmospheric pressure = 1 atmosphere = 1 atm

$$1 \text{ atm} = 1.013 \times 10^5 \text{ Pa}$$
$$= 1.013 \text{ bar}$$
$$= 760 \text{ mm Hg}$$

The bar (barometric) is used in meteorology. The millimeter (mm) of mercury (Hg) is used in medicine.

#2.10.2. EXAMPLES OF PRESSURE.

(1) If you ears are popping, the atmospheric pressure (1 atm) on your eardrums is not balanced by the pressure on the inside of your eardrums. Yawning usually equalizes the pressure.

(2) Assume that you are in outer space in a spherical space capsule of radius 50 meters pressurized to a comfortable 1 atmosphere (atm). Outside the pressure is 0 atm. What is the pressure on the wall of your space capsule? The spherical space capsule must be designed to contain your 1 atm environment and keep you safe.

The surface area A of your spherical capsule wall is:

$$A = (4)\ \pi\, r^2 = (4)\ (3.14159)\ (50\ m)^2\ =\ 3.14159 \times 10^4\ m^2$$

The force F_i pressing outward inside your capsule due to the 1 atm atmosphere is:

$$F = PA = (1\ atm)\ (3.14159 \times 10^4\ m^2\ ) = (1.013 \times 10^5\ N\ m^{-2}\ )(\ 3.14159 \times 10^4\ m^2)$$
$$F = 3.18243 \times 10^9\ N$$

In summary:

Pressure outside your spherical capsule: 0 atm

Outward (positive) pressure P of your spherical capsule: 1 atm

Surface area A of your spherical capsule: $3.14159 \times 10^4\ m^2$

Outward force on walls of your spherical capsule: $+\ 3.18243 \times 10^9\ N$

If your space capsule fails, you must act quickly to pressurize your body to 1 atm (in 30 seconds or so) or you will die from lack of oxygen or you body will be seriously damaged perhaps fatally. Fortunately, the human body is surprisingly tough and if you pressurize quickly there may be no lasting damage.

(3) If your capsule were at the bottom of or under the ocean, then the force F (negative pressure) of the heavy water would press inward trying to collapse your capsule or diving suit. Your body is also surprisingly tough in this case and can withstand a pressure of perhaps 300 meters for a short time if your diving sphere or suit were to fail. But don't try it! Even if you continue to get air, this pressured air will significantly increase the amount of nitrogen in your blood due the higher partial pressure of nitrogen to oxygen (2 to 1) so you would have to rise to the surface very slowly to avoid the bends. Consult medical literature for more specific and authoritative information about the body's ability to withstand high or low pressure.

(4) That being said, the purpose of discussing pressure is to relate it to how the Big Bang created our Universe.

The Space Capsule in previous example (2) was pressurized to an outward (positive) pressure of 1 atm. A balloon is pressurized to a positive pressure of more than 1 atm as you blow it up. You could use the blown-up balloon do Work, if you slowly release the air in the balloon and cause a fan to turn and raise a weight. If you continue to blow up your balloon, it will eventually rupture with a loud "pop" as the air in the balloon rushes out in all directions to equalize pressure with its surroundings.

If the space capsule walls were made thousands of times stronger, the capsule internal positive pressure could be increased thousands of times before the capsule would explode (rupture) and spray its contents and walls perhaps miles away around the capsule as pressure was equalized with the outside pressure.

What if the walls of capsule sitting on the Earth were made trillions and trillions and trillions of times stronger and pressure inside the capsule was increased until it ruptured and sprayed its walls and contents out around the capsule? Perhaps there would be enough positive pressure and energy in the air rushing out to knock the Earth out of its orbit and send it crashing into the sun.

#2.10.3. POSITIVE ENERGY PRESSURE INSIDE BIG BANG SPHERE AND NEGATIVE ENERGY PRESSURE OF EMPTY SPACE OUTSIDE BIG BANG SPHERE ARE KEY TO CREATING OUR UNIVERSE.

A t the time of the Big Bang positive and negative pressures and positive and negative gravities created our Universe. How this happened to create our Universe is described in **section 6.2.** But to better understand the explanations there, the basics of positive and negative pressures and positive and negative gravities are explained below using thought experiments. Physicists love thought experiments as they are much cheaper and faster to run than real experiments.

- To begin, obtain a very large sphere (almost infinite in size). This is about the size of infinite empty space. Pressurize the large sphere to 1 atm. That's the pressure around you right now.
- Pressurize a small sphere of 1 m (meter) to 1 atm.
- Place the small sphere inside the large sphere. The positive pressure pushing out in small sphere is 1 atm. The positive pressure pushing in on the small sphere is 1 atm. The net pressure on the small sphere is 1 -1 = 0 atm.
 - Weigh the small sphere. It weighs for example 20 kg.

- Now pressurize the large sphere to 1000 atm. The net positive pressure on the small sphere inside the large sphere is now 1000 - 1 = 999 atm. Because of the net **positive pressure** pushing in on the small sphere and pressure is energy, the small sphere now has more positive gravity and weighs more. The exact amount more it weighs, however, is miniscule and can be calculated using Einstein's mass-energy equivalence equation (**section 1.2.1**).

- Pressurize the large sphere to 10^{10} atm. The net **positive** pressure pushing in on the small sphere is now enormous $10^{10} - 1 \cong 10^{10}$ atm. Because the huge net positive pressure pushing in on the small sphere is energy, and energy is equivalent to mass, the small sphere now has enormous positive gravity and weighs a lot more.

Now reverse the process so that the pressure in the small sphere is much greater than that of the large sphere. This is much more interesting and informative.

- Pressurize the large sphere to 0.1 (10^{-1}) atm.
- Pressurize the 1 m small sphere inside the large sphere to 1 atm.
- The positive pressure pushing out in small sphere is 1 atm. The negative pressure pushing in on small sphere is 0.1 atm. The net **negative** pressure pulling out on the small sphere is 0.1 -1 = -0.9 atm. Because of the net **negative** pressure pulling out on the small sphere, it weighs slightly less than 2 kg. The large sphere itself now is a source of **negative pressure, negative** energy, and **negative** gravity with respect to the small sphere.

- Now reduce the pressure of the large sphere to 10^{-10} atm (almost zero).
- Pressurize the 1 m small sphere inside the large sphere to 10^{1000} atm.
The net **negative** pressure pulling out on the small sphere is now 10^{-10} atm - 10^{1000} atm $\cong$ - 10^{1000} atm. Because of the huge net **negative** pressure pulling out on the small sphere and **negative** pressure is **negative** energy and **negative** gravity, the small sphere now has is subjected to huge **negative** pressure, huge **negative** energy, and huge **negative** gravity and weighs a lot less.

This is a good "first approximation" of the situation just before the Big Bang that created our Universe out of a highly pressurized small Big Bang sphere into the infinite expanse of unpressurized negative gravity of empty space. There is much more about this in section 6 including the role that temperature plays in creating our Universe.

2.10.3. POSITIVE ENERGY PRESSURE INSIDE BIG BANG SPHERE AND NEGATIVE ENERGY PRESSURE OF EMPTY SPACE OUTSIDE BIG BANG SPHERE ARE KEY TO CREATING OUR UNIVERSE (CONTINUED).

Follow the almost infinite **negative** pressure **and negative** energy of the infinite large sphere:

- It took energy to pump the air out. That negative energy still resides in the infinite large sphere as **negative** energy. If a small sphere pressurized to 10^{1000} atm were inside the large sphere and a valve was opened to release the pressure in the small sphere, all the positive pressure in the small sphere would be rapidly sucked out down to the almost the last particle by the negative pressure and gravity of the large sphere.

Negative pressure and negative gravity seem counter intuitive. But think of the small sphere pressurized to 1 atm and the infinite large sphere pressurized to seemingly zero pressure. Open a valve between them and follow the positive pressure (energy) - the contents of the small sphere will immediately be "sucked" out into the large sphere. Since the large sphere is almost infinite, no difference in its energy or pressure in it can be detected. However, the energy of the pressure from the small sphere is in the infinite large sphere somewhere.   Now bring up another infinite sphere and connect it to the first one that just had the positive pressure from the small sphere "drained" into it. The positive pressure would now drain to the second large infinite sphere and equalize the energy between them. The positive energy has been found!

In any interaction, physicists and other scientists "follow the energy." Following the energy has led to countless thousands upon thousands of discoveries in many different scientific fields. Follow the energy here: The one atmosphere of energy has been entirely transferred to infinite empty space of the large sphere. This transferred energy is available if you can find a way to get at it and use it. One way is to be in space without a space suit. The negative pressure (negative energy) of empty space will immediate act over all your body like a suction cup causing your death in a very short time (if you don't die from lack of oxygen first).

#3.0. OVERVIEW OF WAVE MOTION AND LIGHT CHARACTERISTICS.

In the early 17th century, light was mostly considered to consist of a beam of corpuscles - the corpuscular theory of light. Light corpuscles were considered to travel out from their source in straight lines. They penetrated and were bent (refracted) by transparent materials such as water. They were reflected by opaque materials. Glass lenses (spectacles) made of glass could be used to magnify objects. Curved mirrors could likewise magnify reflected images.

In 1670, Christian Huygens explained the laws of reflection and refraction of light using geometric wave theory (physical optics). However, sound waves were able to travel around corners, but it was believed that light could not, so acceptance of a wave theory was limited. (Later experiments did demonstrate that light exhibits diffraction characteristics and indeed does bend around the edges of an object.)

In the early eighteenth century, experiments on light velocity, interference, and diffraction could only be explained if light is a wave. By the beginning of the 19th century, light was considered by most to be wave.   Its wave properties had been studied in detail including radiation, transmission, velocity, interference, reflection, refraction, and diffraction. Although light demonstrated particle (corpuscle) properties, it was believed that they could all be explained by wave theory.

On the other hand, the electron was considered to be a particle. The charge of an electron had been isolated and measured by Robert Andrews Millikan for which he received the Nobel Prize in 1923. In his oil drop experiment, Millikan had shown that the charge of an electron was negative and could not be divided below a minimum value of that of a single electron.

In the 19th century, the characteristics of wave motion had been extensively investigated including of sound waves, vibrating strings, water waves, vibrating tuning forks, and the like. As the phenomena of light were investigated, scientists concluded that light was also a wave and exhibited many characteristics of waves that are of great interest in this book.   The term light in this book is used for visible and non-visible electromagnetic radiation.

Light could be reflected. It could be refracted. Light waves interfered with each other cancelling and reinforcing.

Water and sound waves require a medium to travel through, so to explain light traveling though air and a vacuum, physicists in the 19th century postulated the existence of a medium for the light to travel through –a solid all pervasive luminiferous ether (aether) – a misconception that was later corrected.   The following discussion in this **section** provides an overview of wave motion in general and is applicable to the wave characteristics of light.

#3.1. OSCILLATING SINUSOIDAL WAVES TRAVEL OUTWARD FROM THEIR SOURCE.

 Drop a stone in a pond or shake one end of a rope. Watch the oscillating sinusoidal peaks and valleys of the waves travel outward from where the stone was dropped or where the rope was held.   Sound, electromagnetic, and other waves also travel outward from their source. The energy of a wave many travel out in an ever-expanding circle as in the case of water waves or an ever-expanding sphere as in the case of sound waves. Various means (reflectors, directors, speakers, horns, antennas, wave guides, and the like) may be used to direct the energy of the oscillating waves into highly directional patterns.

A wave can move an object forward that is moving up and down in it. A wave trough will move an object in it along as it bobs up and down. The troughs (ripples) spread out from the wave origin and carry an object along with it as the wave front advances.

#3.2. TRANSVERSE AND LONGITUDINAL WAVES.

The up and down (oscillatory) motion of the water, rope, and sound waves is perpendicular (**transverse**) to the outward motion of the waves. Waves have their oscillations parallel to their outward path are longitudinal waves. Light is a transverse wave; sound is a longitudinal wave, water waves are transverse waves. Transverse waves can be polarized (**section 3.12**), longitudinal waves cannot.

#3.3. AMPLITUDE, FREQUENCY, WAVELENGTH, PERIOD, PHASE, AND VELOCITY OF WAVES.

Watch as the waves in water pass a single point in the water. The maximum height of the wave above the water is its amplitude A. The rate at which a wave goes up and down is its frequency f.

Frequency is measured in cycles per second (Hertz). The time between successive peaks is the period P of the wave. The distance between peaks is its wavelength λ. The speed that a peak moves along is its velocity v. These parameters are related as follows:

$$P = 1/f$$
$$\lambda = v/f$$

#3.4. PHASE OF WAVE IS MEASURED IN 360 DEGREES.

Wave motion is periodic with recurring cycles. Each cycle is sinusoidal with **amplitude** that varies throughout - a **recurring cycle of zero to 360 degrees**. In each **cycle** the amplitude varies from zero at (zero degrees), to a positive peak (at 90 degrees), to zero again (at180 degrees), to a negative valley (at 270 degrees), and to zero (at 360 degrees) again. The cycle repeats over and over with each new cycle beginning at 0 degrees of the previous cycle. A specific location in a cycle is specified in degrees, for example 102.4 degrees.

Two waves of the same frequency arriving at the same point at the same time may be:
- Exactly in phase and additive. A 5 cm wave and a 10 cm wave would combine to a 15 cm wave.
- 180 degrees and cancelling. A 5 cm wave and a 10 cm wave would cancel to a 5 cm wave.
- Any other relative phases of waves (perhaps one wave at zero degrees when another is at 44 degrees) of the same frequency would combine into a new wave of the same frequency of new amplitude and phase by point to point addition.

See **section 3.9** for more information about **wave interference**.

#3.5. WAVE VELOCITY.

There are many types of waves (sound, water, electromagnetic, vibrating string, etc.).Each type of wave has its own physical origin. Each type of wave travels out from its source at a velocity and character dependent on the medium and the type of wave. Each type of wave has a characteristic velocity.

3.5.1 SOUND VELOCITY.

Sound is a coordinated disturbance of large numbers of molecules. The energy of a sound wave makes molecules of its medium move and collide but not appreciatively change the individual position of each molecule as the wave passes. This can be easily seen as a wave moves along on the surface of water. An individual leaf of piece of wood will bob up and down and only very slowly move along in the same direction as the rapidly passing wave.

Sound requires a medium in which to propagate. Sound does not propagate in a vacuum. Sound travels at a unique velocity and character (transverse or longitudinal) that is dependent on whether it is propagated in air, water, or solid.   Its velocity also varies dependent on temperature and pressure. The velocity of sound waves is independent of the velocity of the velocity of the source of the waves.

Sound Velocity In Air at 0 Deg. C (68 deg F):
        344 meters/sec,   1236 km/hour,   768 miles per hour

Sound travels about 4.3 times faster in water than air, and 15 times faster in iron.

Once sound has entered a medium, its velocity is dependent on the medium it is traveling in and not the velocity of the source.

3.5.2. ELECTROMAGNETIC WAVE VELOCITY.

Electromagnetic waves (**section 1.2.12**) including light waves are unique as, unlike other waves, they do not require a medium in which to propagate and also propagate in many media (air, water, glass, and other "transparent" media). The term "light" as used in this book is intended to encompass the complete range of the electromagnetic spectrum - any restrictions are stated the relevant text.

Light travels faster than anything else. Nothing can travel faster than light which travels 299,792,458 meters per second in a vacuum. Light travels only very slightly slower in air. Light waves travel slower in water than in a vacuum or air.

Unlike other waves, the energy of a light wave is carried by the stream of its particles (or corpuscles as they were called by earlier scientists). We now call light particles "photons" (**section 9**). Light photons to can thought of as being "shot" from a gun such as a light bulb, flashlight, or laser or the sun.

If a gun is mounted on a platform such as a train or car, the velocity v of its bullet is the combination the velocity of the moving platform v_m added to its velocity v_s when fired from the stationary platform: $v = v_s + v_m$

Unexpectedly and astonishingly, early measurements of the velocity of light found that no matter what the velocity of the platform "firing" the light bullets (photons), and no matter what the velocity of the receiving platforms (toward or away from the firing platform, the velocity of light photons (waves) was always measured to be the same. It took Einstein in 1905 to fully explain this strange phenomenon (**section 21**).

#3.6. FUNDAMENTAL FREQUENCY AND HARMONICS OF WAVES.

As noted above in **section 3.3,** the frequency of a wave is measured in cycles per second (cps) or Hertz. Multiples of a fundamental frequency are called harmonics. Submultiples are called sub-harmonics. A fundamental frequency of 256 cps (middle would have harmonics of 512, 768, 1024, … cps; it would have sub-harmonics of 128, 64, 128/3, 42, … cps.

By varying the phase and amplitude of harmonics and sub-harmonics unusual waveforms can be created. For example, the fundamental and an infinite number of odd harmonics result in a rectangular-shaped "square wave."

#3.6.1 DOPPLER EFFECT.

All types of waves exhibit the Doppler Effect. As the source of a wave approaches, more waves arrive than are emitted by the source. As the source recedes, fewer waves arrive than are emitted by the source. The same effect happens if the source is stationary and the recipient is either moving toward or away from a stationary source, or if both the source and recipient are in motion.

Compared to a stationary train, as a train approaches at constant speed, the pitch (frequency) of its whistle is higher. As the train recedes at constant speed, the pitch is lower. The faster the train is traveling toward the source, the higher the frequency at the listener. The faster the train is receding, the lower the frequency at the listener.

A similar situation occurs for light from a distance star. If a star is approaching the Earth, the higher the frequency of its light – a shift toward the violet. If the star is receding, the lower the frequency of its light – a shift toward the red.

Using the definitions in **section 3.3** for wavelength, frequency, and period of a wave, and the **velocity of sound in air as 344 meters per second**, the follow examples of the Doppler Effect in air for a siren are readily calculated:

STATIONARY SIREN 1000 Hz:
      APPROACHING AT 15 M/SEC: 1044 Hz
      RECEDING AT 15 M/SEC: 956 Hz

MOVING SIREN 1000 Hz:
      APPROACHING AT 15 M/SEC: 1046 Hz
      RECEDING AT 15 M/SEC: 958 Hz

IMPORTANT: Note that the velocity of the moving siren makes a larger effect for both approaching (1046 vs. 1044 Hz) and receding (958 vs. 956 Hz) cases. This occurs because the velocity of the moving source must be added to or subtracted from the velocity of sound in air. For electromagnetic waves, this does not happen as the velocity of light is the same regardless of the velocity of the source of the light. This phenomenon led Einstein to his Special Theory of Relativity **(section 21)**.

#3.7. REFLECTION OF WAVES.

Waves (sound, light, water, and otherwise) are reflected when they strike a surface. In general, a wave is reflected from a surface with the angle of incidence equal to the angle of reflection. These angles are with respect with a line that is normal (perpendicular) to the reflecting surface at that point. This can be readily demonstrated for light with any mirror. If the mirror is curved instead of flat, the reflected images may be magnified or reduced – such mirrors are located at carnivals or amusement parks. They are also used in telescopes.

3.8. REFRACTION OF WAVES.

Waves (sound, light, and otherwise) are refracted (bent) when they are travelling in one medium (air for instance) and enter another medium (water or glass for instance) in which they travel slower or faster. The result is that the wave is bent (refracted) as it enters the new medium. The extent of the bending is determined by Snell's Law (attributed to Willebrord Snell in 1621).

$$n_1 \sin \phi_1 = n_2 \sin \phi_2$$

Where n_1 and n_2 are indexes of refraction (velocity) of the incident and refractive media and ϕ_1 and ϕ_2 are the angles of incidence and refraction.

The phenomenon of refraction is readily seen by inserting a straight stick into a pond. It will appear to bend at its entry point into the water.

#3.9. INTERFERENCE - WAVES CANCEL AND REINFORCE EACH OTHER AND THEMSELVES.

If two or more waves meet at the same location, they will reinforce or cancel. Drop two pebbles into a pond. Watch how they create patterns of reinforcement where the waves peak or almost peak at the same time or cancel or almost cancel. Notice as well as the waves proceed on, they continue as if no collision occurred. In rare cases, many ocean waves can come together at the same time and reinforce creating giant "killer" waves.

If a water wave is reflected off the side of a swimming pool, the incident and reflected waves will interfere with each other reinforcing and canceling as they collide. What, you might ask. If light is a wave, why doesn't reflected light interfere with its self? The answer is explained in detail **in the next section** that the incident and reflected light are not coherent and fortunately for us, do not interact with each other or we couldn't see ourselves in a mirror. Light particles doe not readily interact with other light particles.

#3.10. COHERENT WAVES.

As described in the previous **section**, when two or more waves cross paths, they interfere with each other. The velocity of any of the waves may not be impeded, but their energies may add or cancel wholly or partially at any point in position and time depending on their relative phases and amplitudes.

Very interesting interference patterns can be created depending on the relative frequencies and phases of the two (or more) wave trains. These figures from two different wave trains can be made into a display of two perpendicular harmonic motions calla a Lissajous figure named after Jules Lissajous who devised them in 1875.

For the interference patterns to be sensible though, the wave trains must be from coherent stable sources. If one or both of the wave trains jitters in frequency or phase, the resultant interference pattern will be chaotic as well.

Most light comes from the light emitted by atoms as they cool – the sun, electric lights, candles, and so on. These light sources provide short bursts at random intervals and cannot be used a coherent sources. This is why reflected light from a mirror does not interfere with itself. On top of that, light in general does not readily interfere with other light. If it were to readily interfere, everything would be essential dark and we could not see anything

 Ideally for coherent wave trains, the light should come from the same source. Light emitted through a color filter that exits through a small pin hole, is nearly coherent and will produce interference patterns such as seen in the double slit experiment described in **section 7.**

To produce two wave trains of coherent light, a mirror that is also translucent can be used to split the wave train from a single source into two coherent sources. This technique was used to measure the velocity of light by detecting the amount of phase shift of coherent light that traveled two carefully measured distances.

#3.11. PERIODIC WAVES AND BEAT FREQUENCIES (NOTES).

Pluck the string of a guitar or strike a tuning fork. The oscillations continue, cause the air to vibrate, and are heard as a single frequency note and **periodic** (continuing) frequency. In tuning a guitar, musicians and piano tuners use a reference frequency (note) from a tuning fork (or another string or instrument) played at the same time to produce a "beat" note. When two notes are produced together, your ear hears the two notes and the non-linearity of your ear produces the two original notes and also two more notes - the difference and sum of the frequencies of the two notes. The difference note is the **beat note.** The tuner or musician then tunes the piano or other instrument to eliminate the beat note.

The concepts of wavelength and frequency **(section 3.3)** and harmonics **(section 3.6)** are very important in physics. Waves and their harmonics must have exact wavelengths (frequencies) in many instances to continue to exist (fit) and vibrate such as waves on the strings of a guitar or piano or prongs of a tuning fork – other frequencies and wavelengths quickly dampen themselves out. This phenomenon extends to many areas at the quantum level as well, for example, the discrete energy states (wavelengths) electrons can have as they orbit the nucleus of an atom **(section 12.3).**

#3.12. POLARIZED LIGHT.

Transverse waves have the property of polarization. Longitudinal waves do not. A taunt guitar string can be plucked to vibrate vertically or horizontally or anywhere in between. Visible light typically is from billions of random sources that are polarized in all directions (vertically, horizontally, or anywhere in between). A single light "wave" is polarized in one direction.

Waves can be made to be polarized in a specific direction.
- If a vibrating string is threaded through a rectangular slot, it will be restricted to vibrate only in the direction permitted by the slot and will be polarized. It would then not produce a beat note with another string that is vibrating perpendicular to it.
- Light waves that are reflected off a horizontal surface will become more horizontally polarized.
- Radio waves from a vertical AM radio station are vertically polarized. Waves from a horizontal FM or TV station are horizontally polarized.
- Polaroid glass (consisting of parallel crystals) only passes light that is polarized in the direction permitted by the crystals. Polarized glass is used in sunglasses and many other applications.

Light can also be circularly polarized clockwise or counterclockwise.

#4.0. LIGHT RADIATION AND SPECTRUM.

In the late nineteenth century, optical experiments using lenses, prisms, diffraction gratings, and the like enabled extremely accurate investigation of the spectrums of light radiated from various sources at various temperatures [**Roberson**]. Every luminous substance emits a spectrum of radiation that is characteristic of the substance.

Elements and atoms take on or radiate or energy (photons) in discrete steps. Their constituent electrons absorb photons and "jump" to higher energy orbits or emit photons and fall to lower energy states. The change in energy of each orbital electron is in discrete allowable (quantized) steps. If the energy of a photon is more than allowed (required), the excess energy is radiated way as a lower-energy photon. The radiation or absorption of energy is detected as the spectrum of the material being investigated.

However, the same substance may give a different spectrum when heated to luminosity by different means.

Some of these are:
- Introducing the material into the flame of a Bunsen burner.
- Introducing the material into the electric arc between two carbon rods.
- Enclosing a luminous gas or vapor in an evacuated and electrified tube.
-Heating a object made of the material to various constant temperatures.

Several methods to obtain a spectrum confirmed for early investigators that light behaved as a wave and was electromagnetic in character:

- The Doppler Effect: As with sound waves, the velocity of an object changes the spectrum (frequency) of light by moving toward or away from the observer. For example, a distant star viewed by a telescope on the Earth thereby shifting the observed spectrum toward the ultraviolet (higher frequency) as the star approaches or toward the red (lower frequency)as the star recedes.

- The Zeeman Effect: A magnetic field affects the spectrum by breaking each single spectrum line into several components.

- The Stark Effect: An electric field affects the spectrum by breaking each single spectrum line into several components.

As the science of spectroscopy matured, the patterns that appeared displayed several types of "anomalies." These include:

- Fine structure: Spectral lines showed a more detailed composition of doublets triplets, and peaks caused by the small magnetic field of a spinning electron in an atom and also by small relativistic effects of its kinetic energy.

- Lamb Shift: A shift in spectral lines due to the spontaneous production of virtual particles (such as virtual electron/positron pairs and virtual photon pairs) in an atom. The mechanism of this was explained by Willis E. Lamb. His explanation was a confirmation of relativistic quantum theory. In a tiny atom, there is nevertheless a "great" amount of empty space between the orbiting electrons and the nucleus. This "empty space" is filled with virtual particles that pop into existence and then annihilate each other (**section 12.7**). The existence of the virtual particles causes perturbations in the spectrum of the atom. Lamb was awarded a Nobel Prize in 1955 for his work.

-Hyperfine Splitting: Similar to Fine Structure above except due to the small magnetic field of a spinning proton(s) in a nucleus.

- Various Unusual (exotic) Atoms **[Griffiths, c. 5]**: Spectrum anomalies are also due to various types of unusual atoms where either the electron is replaced with another negatively charged particle or the proton is replaced with another positively charged particle or both the electron and proton are replaced. These "exotic" elements are usually very short lived, but some live long enough for their spectrum to be captured. Anti-elements can also exist (for very short times) where positively charged particles orbit a negatively charged nucleus.

#4.1. HEATING OBJECTS TO VARIOUS CONSTANT TEMPERATURES.

If any solid substance (body) is heated it will radiate. As it is heated, it will first become warmer and then glow with a red color. If heating is continued, it will become white-hot. As it is heated, its spectrum is continuous at any specific temperature and wavelength rather than consisting of discrete lines. A plot of the spectrum is made of the radiant energy E versus wavelength λ (λ = c/f where c Is the speed of light and f is the frequency of the photon radiation from the body.)

This continuous spectrum is indicative of the composition of the solid. The energy E radiated from a body heated to and held at a constant temperature is a skewed bell-shape band of energy that varies with the wavelength λ of the radiation. A bolometer, thermopile, or other instrument is used to determine the amount of energy at any specific wavelength. The significant part of the radiated energy plotted by the skewed bell-shape curve of radiation ranges in the mostly visible spectrum from about λ = 0.5 x 10^{-6} meters to about λ = 10^{-7} meters. The maximum value of the radiant energy E_{max} increases with temperature of the object and occurs at a wavelength of about 2.5 x 10^{-6} meters. The wavelength of maximum energy E_{max} decreases slightly as temperature and energy of the radiation increases and the color of the radiation shifts accordingly toward the ultraviolet. The radiation intensity from an object increases rapidly with temperature and is about nine times more intense at a temperature of 1600 degrees (λ = 2.0 x 10^{-6} meters) than at 1000 degrees (λ = 2.6 x 10^{-6} meters).

The specific continuous spectrum obtained varies with the composition and surface texture of the object being heated. An expert can identify in many cases the composition of an object from its continuous spectrum as far as its wavelength of maximum radiation and the radiation at any specific wavelength.
However, investigations into the radiation spectrum of bodies heated to a constant temperature including blackbody radiation did not explain experimental results. Applying wave theories of light resulted in catastrophic errors at ultraviolet wave lengths (high frequencies) - the "ultraviolet catastrophe" where wave theories of light predicted infinite energy to be radiated.

When an object is heated by various methods (conduction, convection, or radiation), the object takes on energy which increases the agitation of its molecules and atoms. Among other things, the object increases in size and mass – an object is larger when hot than cold and weighs more.

Although not clearly understood in the late nineteenth century (as Niels Bohr did not propose his one-electron model of the atom until 1913), at the particle level, when an object is heated, photons of heat bombard the electrons in orbits around the nucleus of the atoms. The photons add quanta of energy to the electrons knocking them into higher energy levels "orbits" around the protons and raising the temperature of the object. Because it takes an exact quantity (quantum) of energy to knock an electron into a higher level energy level, a photon must possess at least a certain minimum energy. Any excess energy is emitted from the interaction as another lower energy photon. Photons of the exact quanta are thus absorbed to excite electrons into a higher energy state while excess energy is emitted as photons from the object giving rise to the spectrum of light being emitted from an object absorbing photons of energy.

As an object cools, the process reverses and energy in the form of heat photons is radiated away from the object. In this case, as electrons fall into lower energy levels and photons are emitted to carry the energy away in quanta of exact amounts of the difference of initial (higher) energy levels and the new (lower) the energy levels.

The absorbed photons from a heated objected object and radiated photons from a cooling object provide a signature of the elements and molecules involved and form the basis of the science of spectroscopy.

Substances also change state from solid to liquid to gas as they become hotter and then change back as they cool. (See section 2.9.4.)

#4.2. QUANTUM THEORY OF LIGHT: BLACKBODY RADIATION - ORIGIN OF QUANTUM MECHANICS.

Considering light as a wave however was found not to completely explain certain observations and experiments of blackbody radiation and the photoelectric effect.

Exacting measurements of the spectrum of objects being heated showed that the spectrum of radiation emitted from an object of any material being heated to and held at a constant temperature depended essentially on the temperature of the object being heated and not its material.

In late nineteenth century, the experimental science of spectroscopy was well advanced, but its quantum underpinnings were not understood. The mathematical equations at the time were not satisfactory. (Physicists are never happy unless their theoretical mathematics leads to calculated results that are in close agreement with the experimental results.) For objects heated to and held at a constant temperature, Wien's Radiation Law held only in regions of very short (ultraviolet) wave lengths (1×10^{-6} meters); the Rayleigh Radiation Law (also called Rayleigh Jeans Radiation Law) held for long wavelengths (8×10^{-6} meters) but not at all for short wavelengths and predicted infinite radiation there, hence the origin of the expression "ultraviolet catastrophe." Neither equation closely matched experimental results except for a very limited region of the spectrum.

At that time it was well understood that all objects heated to and held at a constant temperature emitted essentially the same spectrum of light (with slight variations dependent on the material and surface texture of the objects). An ideal emitter that was entirely independent of the material was called a "blackbody radiator." A blackbody can be essentially duplicated by a cavity within a thermos-like object that could be heated to and held at a constant temperature. To eliminate any influence of the material or the walls of the cavity, a very small hole was made in the cavity and the spectrum of the energy emitted out of the hole was analyzed.

Since the cavity was maintained at a constant temperature, it was at equilibrium. The radiation from the walls of the cavity was not due to electrons of the cavity being excited and jumping to higher energy states of heating and absorbing electrons or decaying to lower energy states of cooling and emitting electrons – the walls of the cavity were in thermal equilibrium hence the name blackbody.

At a constant temperature, the small sample of emitted light from the hole in the cavity is analyzed. A plot was then made of the wavelength λ of the spectrum versus energy E at that wavelength.

In 1900, Max Planck (who had studied radiation for many years) published an equation for blackbody and other radiation. His formula very accurately matched experimental data both at short and long wavelengths. Planck's breakthrough equation was based on the theoretical assumptions that interchange of energy in the processes of radiation and absorption takes place only in multiples of discrete energy units (that we now call quanta). Planck postulated that "oscillators" of absorption and emission in a blackbody had two characteristics:

1. In the interchange of energy E, the oscillators emit or absorb energy E only in multiples of quantized packets (quanta) at a frequency f given by: E = nhf   where h is a constant now called Planck's constant and n = 1, 2, 3, 4, …. Planck's constant h = 6.626 x 10^{-27} erg-seconds = 0.626 x 10^{-34} Joule-seconds.

2. Energy at a frequency f exists only in quanta (packets) of 1 hf, 2 hf, 3 hf, ….
As explained above, Planck made use of the concept of electromagnetic "oscillators" in his theoretical analysis and introduced the concept that energy is quantized. Following up on that reasoning, the reader as well as the writer can correctly jump to the conclusions that not only photons, but all harmonic motion is quantized – an oscillating pendulum, a vibrating string, sound waves of a whistle, a photon, and so on. Further, all energy is also quantized. Energy thus cannot be divided indefinitely. Ultimately, a given amount of energy can only be divided into two pieces so many times until indivisible quanta of energy remain. This is the result promised the reader in **section 2.2.**

4.2. QUANTUM THEORY OF LIGHT: BLACKBODY RADIATION - ORIGIN OF QUANTUM MECHANICS (CONTINUED).

Planck's breakthrough formula for the quantized energy E of blackbody radiation at any frequency f (wavelength λ = f/c) is [**Robertson p. 336**]:

$$E_f = (hf^3/c^2)(e^{hNf/RT} -1)^{-1}$$

Where E_f = Energy of the blackbody radiation at frequency f

    c = the velocity of light

    f = frequency of the radiation = c/ λ

    h = Planck's constant = 6.626×10^{-27} erg-seconds.

    λ = wavelength = c/f

    k = N/R

    T = absolute temperature

    R = absolute gas constant

    N = Avogadro's number = 6.02×10^{23} molecules/mole

Planck's work led quickly to the Bohr atom of hydrogen, Einstein's explanation of the photoelectric effect, and ultimately to a postulate that understanding that the entire Universe is quantized above the particle level, at the particle level, and below the particle level (strings – if they exist). Planck was awarded the Nobel Prize in Physics in 1918.

#4.3. PHOTOELECTRIC EFFECT.

As the sciences of electricity, magnetism, electromagnetism, and light progressed, light was found to cause a metal surface to emit electrons - the photoelectric effect. The intensity of the light was not, however, instrumental in this phenomenon. It was found experimentally that unless the frequency of the light was above a threshold that varied depending on the kind of metal, however, no photoelectric emission took place. The wave theory of light had no such prediction or any reasonable explanation.

A related effect (the Edison Effect) had been observed by Thomas Alva Edison who stuck an isolated terminal (anode) inside one of his early evacuated light bulbs isolated from the heated filament (cathode). A battery and ammeter were connected between the anode and cathode. A large current of electrons was observed to flow from the cathode, through the empty gap inside the evacuated light bulb, to the anode, to the ammeter, to the positive terminal of the battery, and out the negative terminal of the battery back to the cathode completing the circuit. If the anode battery voltage was reduced and a separate battery heating the cathode was reduced in voltage to zero, the cathode to anode current almost stopped except for a trickle of electrons that were emitted by the unheated cathode. The emitted negative electrons made the cathode positive so the trickle of current continued to flow through the evacuated bulb to the anode, to the ammeter, to the anode battery, and back to the now "positive" cathode. (Every time a negative electron gains enough energy to be emitted from the cathode, it leaves behind a positive "hole") that is filled by an electron flowing to the anode, to the ammeter, to the battery, and back to the cathode.) The trickle flow of electrons can be entirely stopped by making the anode voltage slightly negative by an amount (stopping voltage) determined by the work function W of the material of the cathode metal.

This was the invention of an early vacuum tube - a diode - that along with triodes, tetrodes, pentodes, and cathode ray tubes (CRT) that made early radios, TV, and computers possible years prior to invention of the transistor.

4.3. PHOTOELECTRIC EFFECT (CONTINUED.)

Early experimenters made an apparatus similar to the above "diode" to examine the effect of light on a metal – the photoelectric effect. Instead of heating the cathode, high intensity light was shined on the cathode. The cathode (emitter) was made of the metal to be tested. Light of varying frequency and intensity was shined on the cathode. Electrons were emitted from the cathode and flowed to the anode (collector), through the ammeter, to the anode (+) terminal of the battery, and back to the cathode.

But strangely, at the stopping voltage, only light at frequencies above a threshold frequency caused electrons to be emitted from the cathode and flow to the anode. It was found that an increase in light intensity at frequencies above the threshold frequency led to a corresponding increase in current (electron flow from cathode to anode). The increase of current flow was due to the increase of the intensity of light shining on the cathode and caused an increase in the number of electrons.

Variations of voltage polarity, voltage amplitude, light frequency, and light amplitude were investigated, but the salient issue was that light frequency was the predominant parameter and this was not explainable by the wave theory of light. Even if the battery voltage is zero, if the metal cathode is illuminated with light above the threshold frequency, a small current of electrons will flow from cathode to the anode to the ammeter as photons will dislodge electrons from the atoms at the cathode surface. These dislodged electrons leave the cathode positive and cause a current to flow from the anode, to the ammeter, and back to the cathode. A negative voltage applied to the anode that stops this flow is also called the "stopping voltage."

In 1905, Albert Einstein, who had been interested in light since early youth, gave an explanation that followed up on Planck's work and postulated that light has particle properties (as well as wave properties). A light particle (photon) must have a threshold of energy in order to "knock" an electron out of an atom of the metal cathode (emitter). This minimum energy is called the Work Function W of the Metal. Unless the incident light photon has this minimum energy, it cannot dislodge an electron.

Einstein postulated that light waves were made up of discrete energy quanta (later named photons by A. H. Compton). The energy E of a light quantum (photon) is dependent on its frequency f:

$E = hf$ where h is Planck's constant.

The intensity (energy) of light is dependent only on the number of quanta in the light beam. The light quanta behave as particles that travel at the speed of light c.

Einstein was awarded the Nobel Prize in physics for his explanation of the photoelectric effect in 1921. This theory was controversial in the physics community that through many different experiments had come to accept without question that light was a wave not a corpuscle. After a number of years, other quantum theories of matter including the Bohr Atom of hydrogen and the Compton Effect of the interaction of photons with electrons finally proved Einstein was correct and led to acceptance of expanded quantum theories of light, energy, and matter and the duality of particle and wave properties. The duality of particle and wave properties will be investigated further in **section 12.3.**

#5. ELECTROMAGNETIC RADIATION AND PHOTONS.

NOTE: The term "light" is used throughout this book to encompass all electromagnetic (photon) radiation whether visible or not. Light has both wave and particle (photon) characteristics. These characteristics are described in detail in later **sections**.

#5.1. VELOCITY, FREQUENCY, AND WAVELENGTH OF LIGHT.

Velocity of light (c) in a vacuum
186,000 miles per second   299,792,000 meters per second

Velocity of light in water, glass, or other medium is slower by a factor dependent on the medium. Light slowing down as it enters another medium at an angle causes it to bend (refract) as in water. This bending is dependent on the wavelength (frequency) of the light and causes the various wavelengths (colors) to separate as with a glass prism. The separated colors (spectrum) are dependent on the source material of the light and can be used to identify the source material.

Frequency (v) of photons (v is the lower-case of Nu of the Greek alphabet.)
Frequency **v** (cycles per second) is expressed in Hertz (Hz) for photons, other electromagnetic energy, and electromagnetic radiation.

Wavelength of photons (λ)
A wavelength is the distance an **electromagnetic wave** can travel at the velocity of light in one cycle of the wave.

$\lambda = c/v$ (Velocity of light divided by frequency)

#5.2. ELECTROMAGNETIC WAVE SPECTRUM.

Household 60 Hz wall power λ = 3100 miles
AM radio 1000 KHz λ = 982 ft (299 meters)
Channel 2 TV (54 MHz) λ = 18.19 ft (5.54 meters)
Microwave (10^{11} Hz) λ = 2.99×10^{-3} meters = 2.99mm
Visible light (10^{14} Hz) λ = 2.99×10^{-6} meters
X-ray (10^{18} Hz) λ = 2.99×10^{-10} meters
Gamma rays (10^{21} Hz) λ = 2.99×10^{-13} meters

#5.3. ENERGY (E) OF PHOTON.

The greater the photon frequency v or the shorter the photon wavelength λ, the greater the energy **E** of a photon:

$E = hv = hc/\lambda = v(6.626 \times 10^{-27})$ erg-sec
Where h = Planck's constant (See **section 1.2.17.**)

For visible light: $v = 10^{14}$ Hz = 10^{14} /sec
$E = (10^{14}/\text{sec})(6.626 \times 10^{-27})$ erg-sec
   $= 6.626 \times 10^{-13}$ erg

#5.4. EQUIVALENT MASS OF PHOTON.

Photons have an equivalent mass as given by Einstein's formula for the equivalence of mass (m) and energy (E): $E = mc^2$ where c is the velocity of light. $m = E/c^2$
See **section 1.2.1.**

   For visible light, m = $(6.6 \times 10^{-13}$ erg$)/(3 \times 10^{10}$ meter^2/sec^2)
         m = $(6.6 \times 10^{-13}$ gm-cm^2/sec$^2)/(3 \times 10^{12}$ cm^2/sec^2)
         m = 2.2×10^{-25} gm = 2.2×10^{-28} kg

 Although, photon masses are very small, compared to the huge mass of the Sun (2 x 10^{33} gm), the attractive force (F) of gravity is enough to deflect the ordinarily straight path of the photons and bend them toward the sun.
   For two masses, m_1 and m_2, a distance (r) apart , the attractive force (F) of gravity is given by: F= $G(m_1 \times m_2)/r^2$ where G = 6.67×10^{-11} N m^2/kg^2

There is much more to this story though.

As explained in the following **section** and in **section 22**, Einstein's General Theory of Relativity correctly predicts the amount of deflection a photons due to gravitational attraction.

#5.5. DEFLECTION OF PHOTONS BY GRAVITY.

Photons from distant stars are deflected from a straight line toward Earth as they pass near the Sun. This was confirmation of Einstein's General Theory of Relativity and Gravity.

For the Sun and a photon of visible light,

$$F = G_N (m_{sun} \times m_{photon})/r^2$$

$$F = (6.67 \times 10^{-11} \ N \ m^2/kg^2) \times (2 \times 10^{30} \ kg \ \times 2.2 \times 10^{-28} \ kg)/ \ r^2$$

$$F = (6.67 \times 10^{-11} \ N \ m^2/kg^2) \times 4.4 \times 10^2 \ kg^2/r2$$

$$F = 29.35 \times 10^{-9} \ N \ m^2/r^2 = 2.9 \times 10^{-8} \ N \ m^2/r^2$$

The force F between the Sun and a photon is dimensionally in Newtons (N) if r is specified in meters.

However, this value using Newton's equations for the force of gravity gives a result that is only about half of the measured deflection. Several other factors must be considered than just the masses of the sun and the photon being bent. Einstein's equations give an accurate result which consider other factors - the energy-mass including potential energy of the sun and the slowing of light by the sun's huge gravity which bends the photons toward the sun similar to water slowing and bending light in a lake. Einstein's accurate prediction of the bending of light by the sun (**section 22**) was a stunning confirmation of his General Theory of Relativity.

#6. OUR UNIVERSE.

#6.1. COSMOLOGICAL UNITS

Velocity of light $\qquad$ c = 2.99 792 458 x 10^8 m/sec^{-1}

Newtonian Gravitational Constant $\qquad$ G = G_N = 6.6739 x 10^{-11} m^3 kg^{-1} s^{-2}

Light year (ly) $\qquad$ 0.946053 x 10^{16} m

Astronomical Unit (au or A) $\quad$ 1.495 978 707 00 x 10^{11} m

Parsec (pc) $\quad$ 3.0856776 x 10^{16} m = 3.262 ly = 2.06267 x 10^5 A

Diameter of Universe before big Bang $\qquad$ 1 x 10^{-33} cm

Galaxies in Universe $\quad$ 4 x 10^{11}

Black Holes per Galaxy $\quad$ 1, perhaps more. $\quad$ A massive black hole at the center of a $\qquad$ galaxy may be essential for a galaxy to form.

Stars per galaxy $\qquad$ 1 x 10^{12}

Stars in Universe $\qquad$ 4 x 10^{23}

Age of Universe $\qquad$ 13.8 x 10^9 years

Radius of Universe $\qquad$ 25 x 10^9 light-years = 25 x 10^9 (0.946053 x 10^{16} m)

$\qquad$ = (2.365 x 10^{26} m

New stars now being created by typical Starburst galaxy $\qquad$ 2000 per year

New stars now being created by Milky Way galaxy $\qquad$ 2 per year

Volume of Universe $\quad$ = (4/3) (π) (r^3) = (4/3) (3.14159) (6.355314 x 10^{24} m)

$\qquad$ = 1075 x 10^{72} m^3

CMBR temperature $\quad$ 2.7255 K

Stars in Milky Way Galaxy $\qquad$ 300 billion = 3 x 10^{11}

Mass of Milky Way Galaxy $\qquad$ 2x 10^{41} kg

Planets in Milky Way Galaxy $\qquad$ 50 billion

Diameter of Milky Way   6 trillion miles = 6×10^{12} miles = 9.6×10^{12} km

Mass of Universe 3×10^{56} kg

Energy of Universe 10^{50} kg = 9×10^{66} joules

Density of Universe 10^{-23} gm/m^3

Our Solar System, Its Planets, and Pluto formed   4.6 billion years ago

Radius of Sun 6.9551×10^8 m

Mass of Sun (Msolar) = 1 Solar mass = 1.9885×10^{30} kg

Sun Core Temperature 1.5×10^7 K

Sun Surface Temperature 6×10^3 K

Luminosity of Sun (Lsun) 3.828×10^{26} W (watts)

Mean Distance from Sun to Earth   149,597,870,700 m

Mass of Earth 5.9727×10^{24} kg

Radius of Earth 6.378137×10^6 m

- Our Universe was created by a Big Bang at a tiny spherical point.
- Space and time did not exist until our Universe was created
- Time and Space began at the Big Bang

- The further away a galaxy is, the faster it is moving away from other galaxies.
- Since the Big Bang, Space between the galaxies is expanding in all directions at an accelerating rate pushing the galaxies apart at an accelerating rate. Space tells galaxies how to move (**section 22).**
 - Galaxies themselves are not expanding but are moving apart from each other in all directions at an accelerating rate. Black Holes may be preventing galaxies from expanding themselves.

- Everything in our Universe is a descendent from the spherical Big Bang point.

- Every Galaxy seems to be at the center of our Universe including our Galaxy, the Milky Way!
- Our Planet Earth and our galaxy, the Milky Way, appear to be at the Center or the Universe. So do all other planets and galaxies!!!!

#6.2. TIMELINE, TEMPERATURE, AND PRODUCTS OF BIG BANG THAT CREATED OUR UNIVERSE (Adapted from [Riordan]).

1. 0 sec   ∞? K   Big Bang – Creation of Universe begins

Big Bang Sphere (BBS) has been created in infinite empty space (**section 18**). BBS consists of compressed quantum energy "strings "of about Planck diameter (10^{-33} cm) at almost infinite pressure and temperature (**section 17**).

2. 10^{-50} sec 10^{32} K   Plasma of strings.
3. 10^{-35} sec 10^{28} K   Inflation (also called inflaton)

A 10^{30} or more inflationary expansion of hot plasma in 10^{-35} sec at faster than the velocity of light (**section 6.5**) This inflation was caused by pent up pressure and energy of Big Bang sphere into infinite negative pressure, negative energy, and negative gravity of infinite empty spacetime (**section 2.10.3**). Inflation creates all the energy (and mass) of the Universe from the infinite negative energy of empty space. Rapid expansion of the Universe then slows.

4. 10^{-11} sec 10^{16} K   Transition
5. 10^{-7} sec 2×10^{13} K   Quarks and gluons form from plasma (**section 9**).

Cosmic Microwave background Radiation (CMBR) begins (**section 6.7**).

6. 10^{-5} sec 10^{12} K   Neutrons Form (**section 10**)

Quarks and gluons combine to form electrically neutral neutrons. Some electrically charged protons and electrons also form from quarks and gluons but mainly neutrons.

7. 10^2 sec 10^9 K   Neutrons Decay (**section 13.1.1**)

Neutrons decay into protons and electrons. Protons and electrons form clouds of hydrogen (77 percent), helium (23 percent), and lithium (trace).

8. 10^5 yr 10^3 K   Stars form.
9. 10^8 yr 10^2 K   Black holes form enabling Galaxies to form around them.
10. 10^9 yr 10^2 K   Solar Systems form
11. 10^{10} yr   2.7 K   Today

#6.2.1. AS BIG BANG COOLED, HYDROGEN WAS FORMED.

NOTE: Sections 8 through 13 provide detailed information about elementary particles (photons, neutrinos, quarks, gluons, and electrons) and composite particles (neutrons and protons) discussed in this section.

At the Big Bang (section 18), the particles of the Big Bang traveling outward achieved velocities much greater than the velocity of light. Almost infinite velocity was achieved. This was possible as Space itself was being created and expanding. See Section 6.5.)

This extremely rapid inflation of our Universe was caused by the infinite negative pressure and negative energy of infinite empty space (section 2.10.3). The almost negative pressure and negative energy of empty space were thus converted to the enormous positive energy of the mass and energy of our Universe – a Universe out of nothing! It all took just 10^{-35} sec.

How big was the Big Bang? The Big Bang was just big enough and hot enough to get the process of inflation going. It took less than 10^{-50} sec. (See section 17.)

At the instant of the Big Bang, temperatures were too extreme for anything to exist except formless plasma. In first micro-instants after the Big Bang, the plasma began to cool by radiating away energy as photons that are now seen on the Earth as the Cosmic Microwave Background Radiation (CMBR) described in **section 6.7.**

NOTE: In the discussion below, see sections 9 and 10 for listings and descriptions of elementary and composite particles.

As the plasma continued to cool, quarks were able to form. At an instant later, quarks and gluons created electrically neutral particles: neutrons. The neutrons rapidly decayed into stable particles which have electrical charge: protons (+) and electrons (-), and uncharged neutrinos. (See **section 13.1.1.**) Charged particles: protons (+) and electrons (-) were also produced at lower temperatures. The protons and electrons combined and in effect, converted the huge negative energy of infinite empty space into huge clouds of hydrogen.

The seemingly nondescript neutron that decays in about 15 minutes thus played a major role in the creation of our Universe by being responsible for 80 percent of its hydrogen. The neutron will later play a key role when stars turn into factories of all the elements (atoms) in our universe (**section 6.2.3**).

The huge clouds of hydrogen now were at the mercy of gravity, and condensed into denser and denser clouds creating protostars (**section 6.2.2**) fusing hydrogen atoms into more complex element helium and leading to the formation of stars.

The surface temperature (about 6000 K) of a star such as our sun is not high enough to support nuclear fusion to create helium.   At its center, though, the temperature is about 15 million degrees K, high enough to fuse hydrogen into helium which gives off enormous energy in the process. The surface temperature of most other stars ranges up to 60,000 K. The core temperature of the largest stars ranges up to 10^9 K.

After slowing down due to the effect of gravitational attraction of newly formed elements of the early Universe, our Universe has continued to expand and is now expanding at an accelerated rate perhaps due to the negative energy of empty space beyond our Universe.   Our Universe will now continue to age and expand into cold oblivion.

Woops!! What about anti-matter and anti-particles. As the plasma cooled after the Big Bang, both matter and anti-matter particles were created and quickly annihilated each other. Fortunately for every billion or so anti-matter particles that were created, one more matter particle was created, or our Universe would never have been created. See section 13.2.

#6.2.2. TYPES OF STARS – CREATION OF ELEMENTS IN THE UNIVERSE.

A solar mass is the mass of our sun (Msolar).
Stars range from about 0.2 solar masses to 300 solar masses.

PROTOSTAR. A huge cloud of stellar gas, mainly hydrogen, clumps and collapses due to gravitational attraction. Heat creates nuclear fusion changing hydrogen into helium and radiating neutrinos, photons, and gamma rays. This process accelerates over perhaps 100 million years and creates a star. There are billions and billions of galaxies, each with billions and billions of stars (**section 6.1).**

1 SOLAR MASS (1.9885 x 10^{30} kg). Red Dwarf, Life Time: 10^{10} years. As it dies, it swells up to red giant, then, collapses to bright White Dwarf. Ends up cold cinders after about another 100 million years, Our Sun is about 5 billion years old. In another 5 billion years or so our Sun will swell up, engulf the Earth, shrink to a small bright white dwarf, run out of hydrogen, and then end up as cold cinders.

8 SOLAR MASSES. Becomes Supernova. Collapses electrons into protons and becomes neutron star of 1.4 solar masses, 10 Km radius, gravity = 10^{10} g.

MORE THAN 20 SOLAR MASSES. Explodes into supernova. Becomes neutron star of more than 2 solar masses. Collapses to almost infinite density and almost zero radius. Becomes a giant Black Hole that will suck in other stars and black holes.

DYING STARS ARE FACTORIES. All the atoms on Earth were once part of a dying star as were the atoms of the moons, planets, comets, and meteors in every galaxy.

Small stars that are dying create atoms of few protons – hydrogen, oxygen, carbon, and so forth (fortunately the ones needed for living things) but do not have temperatures high enough to create more complex atoms. Larger stars can create more complex atoms. But only the very rare largest stars have high enough temperatures when they die to create the most complex atoms such as gold and uranium.

#6.2.3. DYING STARS ARE FACTORIES THAT CREATED ALL THE ATOMS IN THE UNIVERSE FROM HYDROGEN CLOUDS.

As a star about the size of our sun dies (runs out of hydrogen), the pressure and heat of its fusion of hydrogen into helium are no longer able to keep it from collapsing. As it dies, it balloons up to a red giant and then collapses again; its core temperature rises greatly and helium in turn is fused into higher order elements such as boron, oxygen, and carbon (thank goodness, as all living things are made of carbon).

Mid-size stars have higher core temperatures, produce heaver elements, such as calcium, chlorine, sodium, and iron. Even including these stars, only about 26 of the natural 92 elements will have been produced.

Finally, the star collapses on itself or explodes, showering its galaxy with its remains, and for heaver stars, becoming supernova and neutron stars.

To produce the remaining 66 heavier elements, including the rare elements such as gold and uranium takes a star of about 20 solar masses or more. The core temperature of such star is 10 billion degrees K or more.

The range and types of elements being produced in the fiery factory of each star is different as the mass and core temperature of each star are different from those of other stars. A star's radiation (light) spectrum as detected by a telescope displays exactly what elements are being produced in each star by the colors and absorption lines of its spectrum (**section 4**).

Ultimately, of course, in many hundreds of billion years or so from now, all the hydrogen clouds that were formed by the Big Bang will be completely used up and no new stars can form. Then as existing stars die off, the entire Universe will be cold, our ancestors will have used up all their energy sources, and hopefully they will have journeyed to another Universe. See **section 18.**

#6.3. BLACK HOLES AND QUASARS.

As described in **section 6.2.2**, a dying star of more than 20 solar masses explodes into a supernova, and becomes neutron star of more than 2 solar masses. Then it collapses to almost infinite density and almost zero radius and creates a black hole that will suck in other stars and black holes.

- Diameter of a small black hole (of 5 stars) at the center of a galaxy:
    5 miles
- Diameter of a very massive black hole (of 5 million stars) at the center of a galaxy:
    5 million miles
- Temperature of 12-mile diameter black hole: 10^{11} K.
- Most galaxies, perhaps all, have a massive black hole at their center. Black holes may be essential and required by galaxies to be able to form around them.

- Quasars are very, very bright objects at long distances away that were formed with black holes in our early Universe. The amount of red shift of light from a quasar indicates how old it is. A quasar indicates that a galaxy and its associated black hole are being formed. As the black hole is formed, it accelerates matter into it which produces the bright light of the associated quasar.

- As recently reported in the magazine Nature, an especially brilliant quasar (the brightest and biggest quasar known) was recently discovered whose light was produced when the Universe was only 875 million years old. The black hole associated with this quasar contains the mass of 12 billion suns. The black hole associated with this quasar is among the biggest known even though it is among the oldest. About 40 quasars are known that developed within a billion years of the Big Bang.

- Our galaxy (Milky Way) has black hole of over 4 million solar masses at its center.
- Some galaxies have black holes at their center of several billion stars.

- Black holes radiate energy slowly. This is Hawking Radiation named after the author of the theoretical work, Stephen Hawking. Hawking Radiation is due to interaction of a black hole with quantum empty space (**section 12.7**). If one of a pair of virtual particles (particle and anti-particle pairs) is captured by the black hole releasing gravitational energy, its partner may escape as radiation. This radiation results in a loss of mass by the black hole, and eventually the black hole will entirely disappear as radiation.

- Radiation from black holes has not been detected experimentally.

- Black holes have limited life spans based on their mass: One star mass: 10^{67} years; Galactic mass (billions of stars): 10^{97} years; multi-galactic mass: 10^{106} years.

Mass has positive gravitational energy when it is attracting another mass. The closer it is to the other mass, the greater its attraction for the other mass.   The mass also has negative gravitational energy when it is being attracted by another mass. The further away it is, the less it is attracted to the other mass. The negative gravitational energy of mass is much less than the positive gravitational energy (sections 1.2.1 and 2.3.3). Mass will have more negative energy if its diameter becomes smaller. But a mass can never have more negative energy than positive energy! Before the negative energy can become greater than the positive gravitational energy, the mass will collapse to a black hole with positive gravitational energy. Thus, masses such as stars and black holes cannot be created out of nothing (almost empty space), but a universe can (section 17). That is, negative energy can create positive energy but it can't create more positive energy than it has negative energy. Infinite empty space is an infinite reservoir of negative energy (section 2.10.3)

#6.4. OUR UNIVERSE LOOKS THE SAME IN ALL DIRECTIONS.

The Universe looks essentially the same in all directions from Earth. The Earth therefore appears to be near the center of the Universe. (See **sections 6.1 and 6.7**.)

#6.5. OUR UNIVERSE IS EXPANDING AT ACCELERATING RATE. (See section 6.1.)

At the beginning of the Big Bang, the just born Universe rapidly expanded (inflated) at faster than the velocity of light creating all the mass and energy in the in the Universe. "What?" you may ask. Nothing can obtain a velocity that is faster than the velocity of light. There are several ways that that this might occur. Two are described below.

- Many physicists theorize that **nothing** exists (space or time) outside our Universe. As our Universe inflates by the Big Bang, the Universe is "unfurling" like a balloon or "bubble" into **nothing** and can do so faster than the velocity of light creating all the Space, mass, and energy of our Universe. This line of reasoning has led physicists (and cosmologists) to conceptualize bubble universes, parallel universes, membrane universes, and the like. Of course, if nothing (no time and no space) existed before our Universe, then where was the Big Bang created to form a Big Bang to unfurl and create our Universe?

- Another theoretical way that faster than light velocities might be achieved is to assume that the empty space and time (spacetime) do exist everywhere. Further, assume it is just like empty space and time in our Universe now except it was much emptier and much colder without a Universe in it: just about zero degrees Kelvin perhaps, as cold as 10^{-30} K. At this temperature (**section 1.2.23**), perhaps the Higgs fields (**section 13.7**) that create gravity and restrict velocities to less that the speed of light undergo a change of state (**section 2.9.4**). Further, the Higgs fields are "frozen over" and did not impede velocity increases over the velocity of light or velocity increases at all. In this scenario, the Higgs fields are assumed to undergo a "change in state" just as water undergoes a "change in state" as it freezes (**section 2.9.2**). The Higgs fields in this state do not impede the accelerated expansion of the Universe at more than the speed of light into empty spacetime beyond our Universe. Then, as our Universe expanded, it warmed empty space and the Higgs fields began to function normally to slow down expansion of the Universe. Gravity of newly formed matter then slowed expansion to less than the speed of light to its current rate.

Regardless of how our Universe inflated at faster than the velocity of light, Dark Energy (**section 14.3)** or perhaps negative pressure and negative gravity continue to **accelerate** expansion of the Universe today (**section 14.2.2**).

Eventually, no other galaxies will be visible from our galaxy, the Milky Way. Galaxies themselves are not expanding. The galaxies are moving away from each other as the Universe expands. Today, all galaxies including our own Milky Way are moving away for one another in all directions **at an ever accelerating rate**. The further away the galaxy, the faster it is rushing away. This is due to the Universe continuing to expand into infinite empty space beyond the Universe. The temperature there is almost zero degrees Kelvin. At this temperature, the Higgs fields are not resisting acceleration of the galaxies.

In explanation**, assume every galaxy is 1 km away from its nearest neighbors**. Your neighbor, 10 galaxies away, is also 1 km away from its closest neighbors. After a period of time you re-measure and find your nearest neighbors are **1 km more** away from your galaxy and are now 2 km away. You also now find you neighbor that was 10 km from your galaxy and 1 km away from their closest neighbors is now 2 km away from their closest neighbors, but to your surprise, it is now 20 km away from you. The further a galaxy is away from your galaxy, faster it is receding. The diameter of our galaxy (10km) has doubled, its volume has quadrupled, and its density is one-fourth.

Not only are distant galaxies moving away from each other at a greater rate than close galaxies, the rate of expansion of all galaxies is accelerating at an ever increasing rate. Soon (in billions of years), galaxies might be moving away from each other faster than the velocity of light. If so, soon (in many billions of years) our decedents will not be able to view another galaxy in their Universe.

Unless we leave records, our ancestors will not be able to fathom the birth of their Universe from cosmological observations as our scientists have done.

#6.6. AGE OF UNIVERSE.

The latest best estimate places the age of the Universe at three significant figures to be 13.8 billion years. The previous best estimate made in 2012 to "four" significant figures was 13.72 billion years. The current finding was released in March 2013 from data collected by the Planck Space Telescope. Cosmic Microwave Background Radiation (CMBR) was mapped and analyzed (**section 6.7**). The Planck Space Telescope was launched in 2009.

The red shift of light from the CMBR (**section 6.7**) that we are just now receiving and consideration of (1) the time CMBR light took to get to Earth, (2) the continuing expansion of the Universe, and (3) various other factors enable the age of the Universe to be calculated as 13.8 billion light-years.

#6.7. COSMIC MICROWAVE BACKGROUND RADIATION (CMBR).

CMBR is essentially the same in all directions. It is at the edge of the Universe. The radiation seen now occurred about 350,000 years after the Big Bang as the extreme temperatures of the Big Bang cooled to now very near absolute zero temperature (2.7 K). Since the CMBR looks almost the same from the Earth in all directions, it appears that the Earth is somewhere near the center of the Universe. At the Big Bang, every bit of space of our Universe was at one tiny location. This is why today that every galaxy seems (to itself) to be at the center of the Universe and why the CMBR looks the same in all directions to viewers at our galaxy and viewers at every other galaxy.

Cosmic Microwave Background Radiation (CMBR) was created very near the beginning of the Universe when it was too hot for matter to form. The hot plasma of the Big Bang emitted light (microwave photons) now being detected from the fringes of the expanding Universe as the CMBR. As the Universe continued to rapidly expand, the hot plasma making the CMBR cooled and matter began to condense and form.

 You can see CMBR as "snow" on over the air TV. Physicists were searching for the cause of the "noise" in their "radio telescope" and the hunt finally lead to the conclusion that the noise was radiation that took place shortly after the Big Bang.

The enormous energy of the Big Bang and its rapid inflation **section 6.2**) left an imprint on the CMBR that has been observed by scientists of the "Background Imaging of Cosmic Extragalactic Polarization (BICEPT2) Project." Their findings were announced in the _Los Angeles Times_ on March 18, 2014 in an article by Amina Kahn and will be published in _Nature._ This hard to detect imprint of gravitational waves on the CMBR changed the some of its polarization. Working at the South Pole to obtain clearer reception, the researchers were able to verify the change in polarization.
These findings about the CMBR at the South Pole support the theory that an inflationary Big Bang created our Universe. They also support the existence of gravitational waves (section 2.3.3).

#6.8. CURRENT DIAMETER OF UNIVERSE.

Determining the current diameter of Universe assumes that the Earth is about in the center of the Universe **(section 6.4)**.

- Since the Universe is 13.8 billion years old, it has been expanding in all directions for 13.8 billion years.

- The light (radiation) now being received on Earth from the CMBR at the edge of the Universe left there about 13.5 billion light-years ago, and the distance that light has traveled to the Earth is at least 13.5 light-years. The question is, 'How big was the diameter of the Universe when the light we are now seeing left the CMBR?"

- The diameter of the Universe is at least twice that, or at least 27 billion light-years, and a lot larger. (See below.)

- The Universe has been found by the Planck telescope to be expanding about 3 percent slower than thought.

- Since the light now being received from the CMBR left the CMBR 13.5 billion light-years ago, the Universe has continued to expand in all directions since then for another 13.5 billion light-years. The diameter of Universe may have doubled and now be 27 + 27 = 54 light-years, less 3 percent (Universe expansion may be slowing) = 52 light-years.

Various estimates of the current diameter of the Universe range from about 40 to 90 billion light years.

#6.9. COSMIC RAYS.

Cosmic rays are particles from the sun and other solar systems that bombard the Earth. The radiation is 90 percent protons and 10 percent helium nuclei. Most are dissipated in the upper atmosphere. Pions carry away about half the incident energy. Collisions and decays result producing muons, electrons, neutrinos, and photons. Cosmic rays are also used at high altitude laboratories or in space for experiments of high-energy particle interactions. My first introduction to cosmic rays was at the lab at the top of the Physics Building at the University of Colorado in Boulder, Colorado - elevation 6250feet.

For cosmic rays: $E = 10^6$ eV to 10^{20} eV. Energies of cosmic-ray particles are usually between 10 MeV and 10 Gev. Cosmic rays of energies of several hundred Gev have been detected. These are usually very high velocity protons.

Cosmic rays have very short wavelengths and enormous energy.

#7. WAVE MOTION AND INTERFERENCE CHANGE UNDERSTANDING OF QUANTUM PARTICLES.

Combining vibrating waves of light photons (as well as of larger objects) is analogous to the combining of water waves in the ocean or a lake or sound waves in the air or other media. Waves merging reinforce or cancel depending on their frequency and phase. Canceling background noise in sound has become big business. Reinforcing waves are surfers' utopia. Rarely, freak events occur in the ocean when reinforcing waves create super waves such as the massive ocean wave in 2010 that upended a sailing vessel and broke its main mast, ending a 14-year-old girl's attempt to be the youngest female to sail solo around the world. In January 2011, surfers were on a relatively calm Pacific Ocean in California. Suddenly, a gigantic wave appeared out of nowhere killing one surfer and injuring several others.

#7.1. COHERENT WAVES INTERFERE WITH EACH OTHER TO PRODUCE INTERFERENCE PATTERNS.

Study of particle physics usually starts in earnest with Quantum Mechanics. The first **section** of the text usually describes a wave interference pattern of photons (light) passing from a common coherent source through **two rectangular slits** onto a screen or array of detector). When the beam of photons traveled to the screen, the photons evidently interfered with each other (like two waves interfering with each other) producing a wave interference pattern of light and dark lines on the screen showing that light photons indeed had both particle and wave properties. The photons interfered with each other just like sound or water waves can interfere with each other.

You can easily create a simple, one-slit wave interference pattern (perhaps more precisely a diffraction pattern) yourself by looking through the thin slit between two fingers squeezed together that are held up a short distance away from your eye, palm toward your eye about 8 inches or so away from your eye. Look through the tiny slit at any light source out the window or in a lamp. You will see black vertical lines in the very narrow slit between your fingers. Squeeze your two fingers together and move your palm toward and away from your eye to get the most pronounced black lines in the very narrow slit between your fingers.

#7.2. QUANTUM PARTICLES INTERFERE WITH THEMSELVES AND ARE SEEMINGLY AT TWO (OR MORE) PLACES AT ONCE.

The conceptual problem arises with two slits when the experimenter reduces the number of photons and finds that the wave interference pattern remains even when the number of photons is reduced to one at a time. Somehow, it seems that each single photon travels alone **through both slits at once** and its wave properties cause it to interfere with itself producing the same wave-interference pattern as appeared with multiple photons with one at a time photons.

At the quantum (elementary particle) level, any attempts to determine which slit or slits a photon went through, will destroy the wave interference pattern. This happens because in order to find where a photon is located, another photon or photons must be used and this addition will change the experiment setup, change the outcome of the experiment, and destroy the wave-interference pattern on the screen. The term for this phenomenon is the "Uncertainty Principle" (**section 12.4.1**). At the quantum level, any attempt to determine or measure one significant characteristic of a quantum-size particle will require the use of other elementary particles and perturb the particle being investigated to the extent that a paired (closely-related) characteristic) of the particle being investigated will be significantly changed or cannot be precisely measured.

This same interference phenomena and interference patterns for photons have been created in the laboratory for other and much larger particles such as electrons, atoms, and large molecules that are not electromagnetic. The results are the same showing all matter exhibits wave properties (de Broglie waves – **section 12.3**) and displays similar self interference. This effect is particularly pronounced for subatomic particles, small atoms, and small molecules. Further, as demonstrated in **section 12.4.2**, particles may not only be at seeming two places at once, but also may have a probability of being at many or an infinite number of places at once.

****

#PART II. PARTICLE PHYSICS

#8. PARTICLE CHARACTERISTICS.

This is a good place to stop, take a quick look ahead, and review the Particle Listing Tables in **sections 9 and 10**. Check out the lifetimes of all the particles listed. Almost all have lifetimes of much less than a millionth of a second. Out of the 242 particles - 38 elementary particles **(section 9)** and 283 hadrons **(section 10)** - plus an equal number of anti-particles, the handful remaining after a millionth of a second are:

1. n Neutron `   Lifetime = **880.1 seconds (14.668 minutes)**
      (unless it is securely bound to a proton inside an atom)

2. p Proton   Lifetime = ∞ **(Infinity)**

3. e Electron   Lifetime = ∞

4. γ Neutrinos (3)   Lifetime = ∞
      $γ_e$ Electron neutrino
      $γ_μ$ Muon neutrino
      $γ_T$ Tau neutrino

5. ϒ Photon   Lifetime = ∞

6. g Gluons (8)   Lifetime = ∞

The proton, neutron, and electron form all the atoms that make up all of the material our Universe. All the rest of the particles are unstable and quickly decay into other unstable particles which then decay into other unstable particles, and so forth, until they decay into one or more of the above stable particles. Under all circumstances, the total energy, charge, and other parameters are conserved in total (the same before and after the decays). (See **section 12.2.**)

That is not to imply that the above six particles cannot decay into other particles under some circumstances if given enough energy. For instance, two photons if possessing enough energy can decay into an electron and a positron (anti-electron).

The weak force bosons (W^+ W^- Z^0) and the Higgs bosons (H) and gravitons are short lived (**sections 9.4.2 and 9.4.4**) but if bonded with a long-lived particle can have an infinite lifetime. This is also the case with the neutron n, whose lifetime is ordinarily about 15 minutes, but in a long-lived particle such as a hydrogen atom, it may have an infinite life-time. Likewise, if a weak-force boson or Higgs boson is bonded in an atom to give it energy-mass, it may as well have an infinite lifetime. And if a Higgs boson is bonded in a proton, maybe a collider can knock it loose (**section 13.7.10**).

#8.1. GENERAL PARTICLE CHARACTERISTICS:

G   Generation (1, 2, or 3) Generation number is the composite group number in which the elementary particle falls. Particles in Group 1 were generally theorized and/or detected before those in a later group. The characteristics of a particle in one group made it natural to find a somewhat similar particle in a later group which in some cases led to search for and discovery of other particles. There are searches going on for fourth generation quarks, heavy and light bosons including axions, and heavy leptons with promising but not completely definitive results.

IsoSpin   Isospin may be clockwise (+) or counter-clockwise (-). Isospin started out to be a way to describe how a proton and neutron were the same particle and one could be turned into the other by changing the orientation of the particle's isospin. As the evidence became overwhelming that the neutron and proton were distinct particles seemingly made up of three more elementary particles, isospin became to be the number of the various kinds of quarks in a meson and baryon: Isospin can be 0, 1/2, 1, or 3/2.
Isospin = 1/2 (no. of up quarks – no. of anti-up quarks) – (no. down quarks)
        – (no. anti-down quarks)
Isospin has taken on strict mathematical meaning that isospin is conserved in particle interactions. Isospin is not a real physical entity – it is a mathematical quantum property that leads to highly accurate experimental results.

Isospin is associated with the weak force and is also called weak isospin. Isospin is a mathematical property of particles but has little relationship to a physical characteristic. Isospin is calculated [2 Baggott, p. 81] as follows:

Isospin =   ½ No. of up quarks: + _____
             ½ No. of anti-up quarks   - _____
             ½ No. of down quarks   - _____
             ½ No. of anti-down quarks   - _____
             Total isospin   _____

Example: For the neutron: = -1/2
             ½ No. of up quarks (1)   + _1/2_
             ½ No. of anti-up quarks   - _____
             ½ No. of down quarks (2)   - __1__
             ½ No. of anti-down quarks   - _____
             Total isospin   _- 1/2_

Example: For the proton: = +1/2
             ½ No. of up quarks (2)   + __1__
             ½ No. of anti-up quarks   - _____
             ½ No. of down quarks (2)   - __1/2__
             No. of anti-down quarks   - _____
             Total isospin   _+ 1/2_

8.1. GENERAL PARTICLE CHARACTERISTICS (CONTINUED):

J Spin Spin may be clockwise (+) or counter-clockwise (-). All particles have spin. Spin was originally applied to describe angular momentum of an electron spinning on its axis like a gyroscope or top. Spin is a quantum value; spin occurs in discrete (rather than continuous) quantum values. Some particle have spin of half integers (1/2, 3/2, 5/2, ...). Particles with half-integer values of spin greater than 11/2 have not been detected to a certainty. Other particles have integer values of J Spin (0, 1, 2, ...).

J Spin is a property that has little to do with spin. It at first was an attempt to describe an electron attribute in an understandable way. It now is considered as spin up or down (or clockwise or counterclockwise) but is best thought of as just different states of a particle.

J Spin at the particle level occurs in discrete multiples of quantum units of $(h/2\pi)$ where h is Planck's constant. The $(h/2\pi)$ is understood to be involved in calculations but is omitted in listings. To covert spin to units of angular momentum, the value of spin is multiplied by $h/2\pi$ where h is Planck's constant $(6.626 \times 10^{-34}$ kilogram-meters^2/sec). Planck's constant h has the dimensions of joule-seconds which is angular momentum.

See **sections 9 and 10** for a detailed listing of I and J spins.

8.1. GENERAL PARTICLE CHARACTERISTICS (CONTINUED):

L_e Electron number (0 or 1)
L_T Tau number (0 or 1)
LT   Lifetime (seconds)
L_μ Muon number (0 or 1)

M   Mass in MeV $/c^2$ See **sections 1.2.15 and 12.1.**

Q   Electromagnetic Charge. Leptons have Electromagnetic Charge of 0 or ± 1 (+ 1 or -1). Quarks have charge of 0, $\pm\frac{1}{3}$, and $\mp\frac{2}{3}$. Mesons have charge of zero, as the charge of one of its two quarks is always cancelled by the opposite charge of its anti-quark. Baryons have charge of 0, ± 1, or ± 2. Searches are going on for baryons made up of four or leptons and/or quarks, these baryons would have charge of ± 2, ± 3, and ± 4,.

#8.2. ADDITIONAL CHARACTERISTICS OF QUARKS.

b B   Beauty (0 or 1). Also known as Bottom.
c C   Charm (0 or 1)

d D   Downness (0 or 1)
s S   Strangeness (0 or -1) Strange s quarks were named because they seemed to have strange properties
t T   Truth (0 or 1) Also known as Top.
u U   Upness (0 or 1)

Color   r red   g green   b blue

NOTE: Color also applies to color gluons of the strong nuclear force as well as quarks. See **section 9.4.3** for more information.

#9. ELEMENTARY PARTICLES – FERMIONS (LEPTONS AND QUARKS) AND BOSONS (38).

Life used to be easy in particle physics. First there were Fire, Air, Earth, and Water. In the early 1900's there were Electrons, Protons, and Neutrons. They were called "elementary" particles and were thought to be the smallest constituents of matter out of which everything else was made. Life was pretty easy thanks to Niels Bohr and the Bohr Atom in 1913.

The first particle was discovered in 1887 by J. J. Thompson. Thompson measured the charge to mass ratio of the negative cathode rays of a cathode ray tube. He concluded that the cathode rays were negatively charged particles now called electrons. Thompson also measured their velocity to be one-tenth the speed of light which was greater that any velocity previously measured. Later, Dirac in 1927 predicted (unknowingly) an anti-electron, the positron which was observed in 1931.

Now there are 76 "elementary" particles counting 38 elementary particles and an anti-particle for each of these 38. All of these elementary particles have been observed and verified except for the theoretical graviton.

Development of the Standard Model (SM) of Particle Physics was completed for the most part in 1978 and is continually updated. The SM places elementary particles into two categories: Fermions and Bosons.

Fermions are further categorized as Leptons and Quarks. These elementary particles are classified in the SM in three generations. The negatively-charged electron (e^-) is a lepton. The positive-charged proton (p^+) and neutral neutron (n) are composed of quarks. The tiny, sub-microscopic electrons, protons, and neutrons form the atoms which make up the planets, suns, and galaxies of our Universe. Yet, every electron, proton, neutron, and other sub-atomic elementary particle is exactly identical with every other electron, proton, neutron – not just about identical or almost identical but exactly identical and cannot be distinguished in any way from another of the same elementary particle.

Particles interact with each other through the mediating force of other elementary particles called Bosons. Bosons are force mediators classified in the Standard Model (SM) in four boson groups: Electromagnetic Force photons, Strong Nuclear (Color) Force gluons, Weak Nuclear Force bosons, and Gravitational Force (Higgs and Graviton) bosons. Theoretically a graviton transmits the force of gravity, but as yet it has not been detected and is not included in the SM at this time.

All particles in the SM except gravitons have now been detected to a certainty and verified by experiment. The last elementary particle found for which there is experimental as well as theoretical evidence is the Higgs boson. Experimental evidence of the Higgs boson was announced in July 2012 at the new ultra-high-energy particle collider in CERN, Switzerland. The Higgs boson is associated with the force of gravity. A theoretical companion to the Higgs boson and the mediation of gravity is the graviton boson. Despite the successes of the mathematical formulation of the SM, there are many parameters that can only be determined experimentally. The most notable of these are the masses of all the particles and the relative strength of the bosons of the four forces.

The elementary particles can combine in limited ways to form hadrons. Hadrons are composite particles such as protons and neutrons made up of fermions (leptons and quarks) and bosons. There are two types of hadrons: mesons and baryons. Mesons contain two quarks. Baryons contain three quarks.

There are more than 283 hadrons that have been detected Experimentally To a Certainty" (ETAC). Additional hadrons will be detected ETAC in the future. For each hadron, an anti-hadron is possible and some have been detected. The worst insult of all is that the proton and neutron are not elementary particles after all but are hadrons (baryons) made up of three quarks each. The good news is that the electron and photon are still considered to be elementary particles.

This material is also not a historical treatment of the many highly gifted and insightful physicists whose accomplishments have lead to theorizing and detecting the Higgs boson. This is better left to those who carried out these efforts or who where colleagues of those who did. Several of the references in the **Bibliography** are highly recommended for this purpose.

Sections 9 and 10 provide complete listing of particles and their major characteristics. They are a major reference and resource to be used while reading the other **sections.**

#9.1. ELECTROMAGNETICALLY NEUTRAL LEPTONS (NEUTRINOS) (3).

Neutrinos have no charge and are each their own anti-particle. Neutrinos can spontaneously change from one type to another (neutrino oscillations) requiring them to have differing masses. Masses of neutrinos are not precisely known. J = 1/2.

NOTE: Particle characteristics (Q, Le, L_μ, L_T, G, LT, mass, etc.) are listed and explained in section 8.

v_e (Electron neutrino) Q: 0 L_e: 1 L_μ: 0 L_T: 0   G: 1   LT: ∞ M: 1×10^{-6} or less

v_μ (Muon neutrino)   Q: 0 L_e: 0 L_μ: 1 L_T: 0   G: 2   LT: ∞ M: 2×10^{-6} or less

v_T (Tau neutrino)   Q: 0 L_e: 0 L_μ: 0 L_T: 1   G: 3   LT: ∞ M: 3×10^{-6} or less

#9.2. ELECTROMAGNETICALLY CHARGED LEPTONS (3).

Each charged lepton has a distinct anti-lepton. Spin (J) = 1/2 (clockwise or counter-clockwise).

e⁻ (Electron) Q: -1 L_e: 1 L_μ: 0 L_T: 0 G: 1 LT: 10^{26} M: 0.510999

μ⁻ (Muon)   Q: -1 L_e: 0 L_μ: 1 L_T: 0 G: 2 LT: 10^{-6} M: 105.6584

T⁻ (Tau)   Q: -1 L_e: 0 L_μ: 0 L_T: 1 G: 3 LT: 10^{-15} M: 1776.82

#9.3. QUARKS (18).

Each of six quarks listed below can have the Color red, blue, or green, making 18 quarks in all. Each of the 18 quarks has a distinct anti-quark.
Anti-quarks have anti-colors (anti-red, etc.).   Free quarks cannot exist.
Quark spin: J =1/2   I =1/2 (u, d) or 0 (s, c, b, t

Quark	Q	D	U	S	Ch	B	T	G	M	I
u (up)	+2/3	0	1	0	0	0	0	1	2.3*	1/2
c (charmed)	+2/3	0	0	0	1	0	0	2	1275	0
t (top)	+2/3	0	0	0	0	0	1	3	173,500	0
d (down)	-1/3	-1	0	0	0	0	0	1	4.8	1/2
s (strange)	-1/3	0	0	-1	0	0	0	2	95	0
b (bottom)	-1/3	0	0	0	0	-1	0	3	4150	0

*Note: M (mass) for all quarks is shown for bare mass and is a calculation. The effective mass in a meson or baryon differs substantially (section 13.4.5). Masses for mesons (two quarks) and baryons (three quarks) in section 10 are the effective values determined by experiment.

#9.4. BOSONS (14).

Each boson is its own anti-boson.

#9.4.1. ELECTROMAGNETIC FORCE BOSON (PHOTON) (1).

ϒ **(Photon)**   M: 0 (less than 1×10^{-18})   I: 0.1   Spin (J): 1^- LT: ∞

#9.4.2. WEAK NUCLEAR FORCE BOSONS (2). W⁻ is anti-particle for W⁺.

W^+	M: 80,385	Spin (J): 1	LT: 10^{-25}	Q: +1
W^-	M: 80,185	Spin (J): 1	LT: 10^{-25}	Q: -1
(Anti-W⁺ boson)				
Z^0	M: 91,187.6	Spin (J): 1	LT: 10^{-25}	Q: 0

#9.4.3. STRONG NUCLEAR (COLOR) FORCE BOSONS (GLUONS) (8).

g   M: 0   I = 0   Spin (J): 1^- LT: ∞   Q: 0

There are eight strong (color) force gluons .

Each gluon carries a combination of color and anti-color of quark colors red, blue, and green. To simplify, for most purposed, each gluon may be consider to carry a color and its anti-color such as red and anti-red. Section 13.4.1 provides a more accurate information about the colors that each of the eight gluons carry.

#9.4.4. GRAVITATIONAL FORCE BOSONS (3).

H⁻ is anti-particle for H⁺. Graviton has not been observed.

H^0 Higgs 0	M: 126,000*	Spin (J): 0 (*)	LT: 10^{-24} (*)
H^+ Higgs +	M: 126,000*	Spin (J): 0 (*)	LT: 10^{-24} (*)
H^- Higgs -	M: 126,000*	Spin (J): 0 (*)	LT: 10^{-24} (*)
g (graviton)	M: 0 (less than 7×10^{-32})*	Spin (J): 2 (*)	LT: 10^{-24} (*)

*Gravitational bosons were detected in July 1012 and are being extensively investigated at CERN. More accurate values should be available in 2014.

#10. COMPOSITE PARTICLES - HADRONS (MESONS, BARYONS, TETRA-QUARKS, AND PENTA-QUARKS).

NOTE: The reader is urged to utilize the CERN website to obtain the latest information on particle searches and detection. Important discoveries and experiments such as detecting the Higgs boson and penta-quark are immediately posted the CERN website.

Mesons consist of two quarks (a quark and an anti-quark).

Baryons consist of three quarks or three anti-quarks.

Sections 9.3 and 10.1 through 10.3 list and describe the specific quark content of mesons and baryons. All these composite particles are part of the Standard Model (SM) of particle physics (section 9) that was developed initially in the 1950's. All those mesons and baryons listed have been detected Experimentally To A Certainty (ETAC) at CERN and several other locations around the world. The lifetimes of these particles are given in these sections and are extremely short much less than a millionth of a second with only a very few exceptions (section 8).

Tetra-quarks consist of four quarks probably two mesons perhaps rotating around each other. Tetra-quarks were hypothesized in the 1950's as part of the standard model (SM) of particle physics but were not experimentally detected until much later. Two tetra-quarks were experimentally detected over several years early in this century, but the experiments did not offer adequate replication so as to be considered detected ETAC. In April 2014, two tetra-quarks were detected at CERN ETAC. At that time the CERN collider was operating at about 6 TeV. The constitution of the tetra-quark was c $\bar{c}$ d $\bar{u}$ (charm, anti-charm, down, anti-up).   The lifetime of this tetra-quark is less than a millionth of a second.

Penta-quarks consist of five quarks, perhaps one meson and one baryon rotating around each other. Penta-quarks seem to be a natural product of very high-energy particles. On July 14, 2015, CERN announced detection of a penta-quark ETAC using 13 TeV. This is about twice the huge power required to detect the Higgs boson. The constitution of the penta-quark was u u d c $\bar{c}$. The lifetime of the penta-quark is less than a millionth of a second.

What about hexta-quarks (6), hepta-quarks (7), … deca-quarks (10), …? Are these possible? Probably, but at energies well beyond what can be achieved soon.

It may be such complex particles were formed right after the Big Bang and acted to absorb, store, and distribute the huge spike of energy. These complex particles would then act to control the Big Bang and steadily release its energy as the quarks and gluons that formed the protons and neutrons of our Universe (section 6.2).

There may also be a place for top quarks which as yet have not been found in mesons, baryons, or more complex multi-quark particles. The top quark is thousands of times heavier than other quarks yet seems to be almost entirely ignored by nature.

RULES FOR COMBINING QUARKS. Mesons and baryons are colorless. A colorless meson is formed by a quark and anti-quark of a color and the same anti-color. A red quark can only form a meson with an anti-red quark, and so forth. A colorless baryon is formed only with three quarks, each of a different color: a red, a blue, and a green quark, or an anti-red, an anti-blue, and an anti-green quark. These rules apply as well to tetra-quarks and penta-quarks.

NOTE: The _**2012 and 2014 Reviews of Particle Physics**_ (hereinafter referred to as **RPP**) list and describe in detail the elementary particles and various groups of hadrons (mesons and baryons). (See **Bibliography**.) Listings in the **RPP** show elementary particles, mesons and baryons that have been observed and detected to a certainty and also those for which experimental data is insufficient and of uncertain interpretation. The elementary particles, mesons, and baryons specified herein are those that have been observed to a certainty. The **SYMBOL** of many of the mesons and baryons as shown below may have additional superscripts and/or subscripts (such as a_2 and f' beyond those shown in the following examples. If a + superscript is shown, 0 and − superscripts may also exist but may not always be shown; and similarly for other superscripts and subscripts. **LT** specifies the "range (from − to) of lifetimes" of particles in a group. Specific lifetimes, masses, subscripts, superscripts, and other characteristics for all elementary particles and hadrons are specified in the **RPP**.

Specific mesons and baryons in the **RPP** are identified by their symbol. If ambiguous, their mass is added after the symbol, for example: $\omega(782)$ and $\omega(1650)$.

#10.1. EXAMPLE MESONS.

NOTE: For the example mesons below, the example positive π^+, negative ⁻, and neutral π^0 particles are shown. Only the neutral particle (0) or positive particle (+) may be shown for additional mesons and baryons. The positive and negative (anti) versions may also exist with the quark changed to an anti-quark, the anti-quark changed to a quark, polarity changed for the charge (Q), and polarity changed for Major Decay Modes (Decay).

π^+　　Quarks: u$\bar{d}$　Q: 1　SCB: 0 0 0　M: 0139.57018　I: 1　J: 0
LT: 2.6033x10^{-8}　　　Decay: $\mu^+ + v_u$

π^-　　Quarks:: d$\bar{u}$　Q:-1　SCB: 0 0 0　M: 139.57018　I: 1　J: 0
LT: 2.6033x10^{-8}　Decay: $\mu^- + v_u$

π^0　　Quarks: (u$\bar{u}$ - d$\bar{d}$)/$\sqrt{2}$　SCB: 0　　0 0 0　M: 134.9766　I: 1　J: 0
LT: 8.52x10^{-17}　　　Decay: $Y + Y$

K$^+$　　Quarks: u$\bar{s}$　Q:1　SCB: +1 0 0　M: 134.9766　I: 1/2　　J: 0
LT: 8.52x10^{-17}　　　Decay: $e^+ + v_e$

#10.2. MESON LISTINGS: Qty 118 [197].

NOTE: Quantity (Qty) of mesons and hadrons detected ETAC is given from the 2012 *RPP* followed by the quantity from the 2014 *RPP* in brackets if different.

SEE *RPP* FOR LISTINGS OF MASS, DECAY MODES, AND OTHER MESON CHARACTERISTICS THAT ARE NOT INCLUDED BELOW.

Meson Type: **Light Unflavored**   Qty: 47 [76] $S = C = B = 0$ *LT:* 10^{-17} to 10^{-8}
Symbol: $\pi, \rho, a, b, \eta, \omega, f, \phi, h$
Quarks: $u\bar{d}, (u\bar{u} - d\bar{d})/\sqrt{2}, d\bar{u}$ $I = 1$ $J = 0, 1, 2, 3, 4$
      $c_1(u\bar{u} + d\bar{d}) + c_2(s\bar{s})$ $I = 0$ $J = 0, 1, 2, 3, 4$

Meson Type: **Strange** Qty: 15 [28] $S = \pm 1, C = B = 0$ *LT:* 10^{-10} to 10^{-8}
Symbol: K (kaon)   **Quarks:** $u\bar{s}, d\bar{s}, \bar{d}s, \bar{u}s$ $I = \frac{1}{2}$ $J = 0, 1, 2, 3, 4$

Meson Type: **Charmed**   Qty: 8 [15] $C = \pm 1$ *LT:* 10^{-15}
Symbol: D **Quarks:** $c\bar{d}, c\bar{u}, \bar{c}u, \bar{c}d$ $I = \frac{1}{2}$ $J = 0/1/2$

Meson Type: **Charmed, Strange**   Qty: 6 [9] $C = S = \pm 1$ *LT:* 10^{-15}
Symbol: D   **Quarks:** $c\bar{s}, \bar{c}s$ $I = 0$ $J = 0, 1, ?$

Meson Type: **Bottom**   Qty: 7 [11] $B = \pm 1$ *LT:* 10^{-12}
Symbol: B   **Quarks:** $u\bar{b}, d\bar{b}$ $I = \frac{1}{2}$ $J = 0, 1, 2$

Meson Type: **Bottom Strange**   Qty: 4 [5] $B = \pm 1$ $S = \mp 1$ *LT:* 10^{-12}
Symbol: B_S   **Quarks:** $s\bar{b}, \bar{s}b,$ $I = 0$ $J = 0, 1, 2$

Meson Type: **Bottom Charmed**   Qty: 1 $B = C = \pm 1$ *LT:* 10^{-12}
Symbol: B_C   **Quarks:** $c\bar{b}, \bar{c}b$ $I = 0$ $J = 0$

Meson Type: $c\bar{c}$ Qty: 16 [29] *LT:* 10^{-17} to 10^{-8} (?)
Symbol: $\chi, \eta_c, J/\psi, \psi, X, \eta$ **Quarks:** $c\bar{c}$ $I = 0/?$ $J = 0, 1, 2, ?$

Meson Type: $b\bar{b}$ Qty: 14 [22]   *LT:* 10^{-17} to 10^{-8} (?)
Symbol: $\chi, \Upsilon, \eta, h, X$ **Quarks:** $b\bar{b}$ $I = 0$ $J = 0, 1, 2$

TOTAL   Qty: 118 [197] Mesons

#10.3. EXAMPLE BARYONS.

BARYON: N⁺ *QTY:* **1** *SYMBOL:* p, N⁺ (proton) *QUARKS:* uud
 S = 0 **Mass:** 938.272 **LT:**10^{32} I = 1/2 J = 1/2

BARYON: n *QTY:* **1** *SYMBOL:* n, N⁰ (neutron) *QUARKS:* udd
S = 0 **Mass:** 939.565 **LT:** 880.1 I =1/2 J = 1/2

#10.4. BARYON LISTINGS (86).

BARYON: N *QTY:* **17** *SYMBOL:* p, N⁺, n, N⁰ *QUARKS:* uud, udd
S = 0 **LT:** 880.1 and 10^{32} I = 1/2 J = 1/2, 3/2, 5/2, 7/2, 9/2, 11/2
BARYON: Δ *QTY:* **10** *SYMBOL:* Δ *QUARKS:* uuu, uud, udd,ddd
S = 0, **LT:** 10^{-17} to 10^{-8} I = 3/2 J = 1/2, 3/2, 5/2, 7/2, 11/2
BARYON: ^ *QTY:* **14** *SYMBOL:* ^ *QUARKS:* uds
S = -1 272 **LT:** 10^{-10} I = 0 J = 1/2, 3/2, 5/2, 7/2, 9/2
BARYON: Σ *QTY:* **12** *SYMBOL:* Σ *QUARKS:* uus, uds, dds
S = -1 **LT:** 10^{-20} to 10^{-10} I = 1 J = 1/2, 3/2, 5/2, 7/2, 11/2
BARYON: Ξ: *QTY:* **7** *SYMBOL:* Ξ *QUARKS:* uss, dss
 S = -2 **LT:** 10^{-10} I = 1/2 J = 1/2, 3/2, 5/2
BARYON: Ω *QTY:* **2** *SYMBOL:* Ω⁻ *QUARKS:*sss
 S = -3 **LT:** 10^{-15} I = 0/? J = 3/2, ?

BARYON: **Charmed** *QTY:* **19** C= + 1 **LT:** 10^{-15} I = 0 J = 1/2, 3/2, 5/2,?
SYMBOL: ^c **QUARKS:** udc
SYMBOL: Σc **QUARKS:** uuc, udc, ddc
SYMBOL: Ξc: **QUARKS:** usc, dsc
SYMBOL: Ωc: **QUARKS** ssc

BARYON: **Bottom** *QTY:* **5** B = -1 **LT:** 10^{-12} I = 0, 1/2, 1 J = 1/2, 3/2, ?
SYMBOL: ^b **QUARKS:** udb
SYMBOL: Ξ **QUARKS** usb, dsb
SYMBOL: Ω **QUARKS** ssb

TOTAL 86 Baryons
TOTAL 204 [283] Hadrons (Mesons and Baryons)

#11. ELEMENTARY PARTICLE DESCRIPTIONS.

#11.1. FERMIONS – DESCRIPTION.

Fermions are elementary particles of matter. Fermions are smaller than a proton which has a diameter of $1.7536 \, 10^{-13}$ centimeters. Fermions are leptons and quarks. Fermions have half-integer spin (J). No two fermions can share the same quantum state. This is called the Pauli Exclusion Principle.

For each fermion there is an anti-fermion making 48 in all. The three uncharged fermions (neutrinos) act as their own anti-particle. For charged fermions, anti-fermions are identical to fermions except for a different polarity of electrical charge. Anti-fermions are identified by a bar over the particle's symbol. The exception to this rule is for the charged fermions with a charge of -1. For the electron identified e^-, the anti-electron is named the positron and identified e^+. The two other charged anti-fermions also use this scheme and are identified as μ^+ and T^+.

Fermions fall into three generations: G1, G2, G3 (first, second, and third). Each lepton and quark has three particles, one in each generation.

#11.2 . LEPTONS – DESCRIPTION.

Three leptons carry charge and each has a distinct anti-lepton. The other three leptons (neutrinos) are without charge, and are each their own anti-particle.

Leptons have Spin (J). The three anti-leptons have the polarity of electromagnetic charge reversed.

#11.2.1. ELECTROMAGNETICALLY NEUTRAL LEPTONS (NEUTRINOS) – DESCRIPTION.

Neutrinos do not have charge and are not affected by the electromagnetic force. They are also not affected by the strong nuclear force. They have almost no mass and are very little attracted by other masses. Neutrinos are affected by the weak nuclear force and are used in experiments of it. Such experiments are difficult to perform because the neutrinos have so little mass.

Neutrinos travel at near the speed of light. Neutrinos are particles of energy that carry away energy in particle interactions in accordance with the requirements of conservation of energy. For example, when a neutron (charge 0) decays into a proton (charge +1) and an electron (charge -1), charge is conserved (+1 -1 = 0) but the combined mass of a proton and electron is less than the mass of a neutron. The missing mass puzzled physicists until neutrinos were theorized and detected.   Experiments have shown that the three neutrinos may oscillate and change from one of the three types into another of the three types. The electron neutrino v_e is very weakly interacting and does not combine with other particles. It is produced by processes such as nuclear fusion that take place on the sun. Trillions and trillions of electron neutrinos from the sun bombard the earth every second with no effect.

#11.2.2. ELECTROMAGNETICALLY CHARGED LEPTONS - DESCRIPTION.

The electron (e⁻) is the best known lepton; it carries a charge of -1. The positron (e⁺) is the anti-electron; its charge is +1 and its electron number is +1; the other characteristics for the electron and positron are identical.

The other two electromagnetically charged leptons are the μ⁻ (Muon) and T⁻ (Tau). They carry the same charge as the electron. They are much more massive than the electron and decay very rapidly usually into an electron. Atoms have been made using the muon and tau in place of the electron in laboratory experiments, but these pseudo atoms are very unstable and short-lived.

#11.3. QUARKS – DESCRIPTION.

Eighteen (18) Quarks fall into three generations: G1, G2, G3 (first, second, and third). There are also 18 anti-quarks with all signs reversed. Quarks carry color – red, green, or blue. Anti-quarks have anti-colors. Only "colorless" combinations of quarks can exist. A color and its anti-color cancel so a quark and an anti-quark can combine and form a "colorless" meson. Three quarks each of a different color can combine to form a colorless baryon. Three anti-quarks each of a different anti-color can combine to form a colorless anti-baryon. Single quarks cannot exist as only colorless particles can exist. A quark can exist only when combined into a colorless meson or baryon. Further, most quark combinations are very, very short lived – much less than a millionth of a second until they decay. Only two quark combinations, the proton (10^{32}) and the neutron (880.1) have lives of longer than a second (**section 10.3)**.

Each quark comes in one of six flavors: d (down), u (up), s (strange), c (charm), b (bottom), and t (top). In addition, each quark comes in one of three colors (red, green, or blue) making a total of 18 different quarks in three generations. For each of these 18 quarks, there is an anti-quark making a total of 36 different quarks. Quarks have spin J = ½.

Free quarks are never seen. If a quark would be ripped out of a meson, baryon, tetra-quark, or penta-quark, strong color force bosons would immediately combine it and its former mate with two virtual quarks **(section 12.7)**. Thus, for example, two quarks of a meson that are somehow parted, would immediately pair up with two other virtual quarks that are created by the energy of the bosons that were created by the energy of separation, so there would be two mesons and four quarks, and so forth for the other quark composites.

The basic constituents of matter are atoms. Protons and neutrons of atoms are composed of quarks. The proton is formed by two up quarks and one down quark (uud) and has an electrical charge of +1. The neutron is formed by two down quarks and one up quark (udd) and is electrically neutral.

The basic constituents of atoms are nuclei of protons and neutrons orbited by electrons. The number of electrons and protons is the same. The hydrogen atom, for example, has one proton orbited by one electron. The electron neutrino v_e is the only first generation lepton particle that is missing from ordinary atoms. This is because it is so weakly interacting that it is never bound to other lepton particles. Quarks interact with all bosons.

Free (single) quarks are never encountered experimentally and cannot exist. They are always entwined with other quarks as mesons and baryons. If two (or three) quarks were torn apart during an experiment or otherwise, the energy required is so large that each solitary quark would be immediately paired with a virtual quark from "empty space (**section 12.7)** to form another meson or baryon.

Quarks exist as mesons of a quark and an anti-quark or baryons of three quarks or three anti-quarks. The strong nuclear force gluons bind quarks together. There are 18 quarks and 18 anti-quarks. These 36 elementary particles have been identified to exist in at least 283 different detected Experimentally To A Certainty (ETAC) hadron combinations. Some of the experimentally detected combinations are discussed below. More complex combinations of quarks can also exist (**section 10.1**).

#11.4. BOSONS – DESCRIPTION AND RELATIVE STRENGTH.

The interactions between particles is determined by the forces acting on (interacting with) the particles. The four forces are electromagnetic, strong nuclear (color), weak nuclear , and gravitational

The most common forces are the electromagnetic and gravitational forces. These are both "square-law" forces. The force F of attraction between two particles (for gravity or electrical charges of opposite polarity) or repulsion (for electrical charges the same polarity) varies inversely with the square of the distance r between them: $F = C/r^2$ where C is a constant related to the strength of the Force. As the distance r between the particles approaches zero, the Force F approaches infinity asymptotically; as the distance r between the particles approaches infinity, the Force F approaches zero asymptotically. The electromagnetic force is much stronger than the gravitational force – see below.

The strong nuclear (color) force is very different. It displays what is called "asymptotic freedom." It operates at short ranges also. When quarks within a hadron (meson or baryon) are close together they behave as free particles - free at least from the strong nuclear force. As distance between quarks in a meson or baryon increases, the strong nuclear force increases very rapidly and approaches infinity. If a "free" quark would stray and venture "outside" the hadron, the energy of the color force is strong enough to spontaneously create two quarks: an anti-quark to form a meson with the straying quark and another quark to replace it in the hadron from whence the straying quark came.

Every interaction between particles has one or more mediator bosons. For electric, magnetic, and electromagnetic forces, it is the photon γ. For weak nuclear force, it is three bosons: W^+, W^-, and Z^0. For strong nuclear force, it is eight color-force gluons g. For gravitational force, it is theorized that there are the three Higgs bosons H^+, H^-, and H^0 and (perhaps) the as yet undetected graviton boson. The electromagnetic force, weak nuclear force, strong nuclear force, and gravitational force are thought to merge into a single force under extreme conditions of temperature and pressure.

Nevertheless, under the conditions in most places in the Universe, there are wide differences in the relative strength of the four forces. Starting with the weakest and ranking the forces in order by increasing relative strength [Griffiths, p 59]:

FORCE	RELATIVE STRENGTH OF BOSONS	
Gravity	$= 10^{-42}$	H^+, H^-, H^0, and graviton bosons
Weak Nuclear	$= 10^{-13}$	W^+, W^-, and Z^0 bosons
Electromagnetic	$= 10^{-2}$	photons
Strong Nuclear Color	$= 10^{1}$	Eight strong nuclear (color) gluons

Each of the bosons is its own anti-boson. The W boson can act as a W^+ boson and also act as its anti-boson. It is indicated as W^- when it is acting as an anti-boson. Bosons do not obey the Pauli Exclusion Principle.

Gluons (g) do not carry electric charge. Strong Nuclear force gluons carry "color" charge. Gluons can absorb and redistribute color charge. The W bosons can absorb and redistribute electric charge.

The Higgs boson is associated with the force of gravity. It was theoretical until July 2012 when it was reported found at the high-energy collider at CERN in Switzerland.

Photon bosons and strong nuclear (color) force gluon bosons are massless. Weak force bosons and gravitational (Higgs) bosons have relatively large mass.

The large masses of the weak force bosons give them very short de Broglie wavelengths (**section 12.3**). As a result, the weak force has a very short range of influence (10^{-18} meters) deep within an atom or proton which is much less than the radius of the proton (10^{-17}) centimeters. The weak force is thus short range rather than weak. The force of gravity is very weak but is infinite in range. The gravitational force of extremely large masses, however, can be extremely large.

The Higgs fields and Higgs bosons give mass to all particles that carry mass (**sections 9 and 10**). The as yet undetected graviton is theorized to provide the force of gravitational attraction, but perhaps, the warpage of spacetime will explain this phenomenon if the graviton is not found to exist (**section 13.7**).

11.4. BOSONS – DESCRIPTION AND RELATIVE STRENGTH (CONTINUED).

The force of interaction between particles is generated by the exchange of bosons appropriate for the force being mediated between particles that carry the charge:
- Electromagnetic force photons for the electric charge (+ or -).
- Weak nuclear force bosons for weak isospin. (Weak isospin applies to leptons and quarks.)
- Strong nuclear force gluons for color charge of quarks, mesons, and baryons.
- Gravitational force bosons (and perhaps gravitons) for mass (force of gravity) for all particles that carry mass

Every interaction between particles has a mediator. As described in **sections 13.3 thru 13.3.3**, two protons or two electrons that have the same electric charge (+) "know" that they are suppose to interact by repelling each other because mediator photon waves (bosons) are being passed between them pushing them apart. Similarly, a proton (+) and an electron (-) have different electrical charge and "know" that they are suppose to attract each other because mediator photon waves are pushing (pulling) them together.

While all this is going on, particles that have mass are being pulled together and being pulled toward other particles with mass by the Higgs fields and the warping of spacetime mediated by the Higgs bosons and possibly gravitons. In most cases, the photons win, as the electromagnetic force transmitted by photons is much stronger than gravitational attraction except where enormous mass is involved.

A very similar process takes place for the strong nuclear force interactions of the eight color charge gluons. For the weak nuclear force, W and Z^0 bosons do the job of mediating between particles.

More detailed information about bosons and boson interactions with other particles is provided in **section 13**.

#11.5. HADRONS (MESONS, BOSONS, AND QUARK-COMPOSITES) - DESCRIPTION.

There are several hundred hadrons of quark-composites (**section 10**) that have been discovered (detected Experimentally To A Certainty- ETAC) – more will be discovered. For every hadron, an anti-hadron is possible. Anti-hadrons are made up of anti-particles. Most hadrons, except the neutron and proton are very short lived, much less than a millionth of a second. Then they decay further until a stable particle is formed.

Hadrons are composite particles made up of color-neutral combinations of quarks and bosons. The two types of hadrons are mesons and baryons. Quarks making up mesons and baryons are held together by Strong Nuclear (Color) Force Bosons (Gluons).

Hadrons also have corresponding anti-hadrons. For example, a proton consisting of three quarks has an anti-proton consisting of three counterpart anti-quarks. Putting a proton together with three quarks (uud) and an electron creates a hydrogen atom. Putting an anti-proton together with an anti-electron (a positron) and anti-bosons will produce anti-hydrogen.

#11.6. MESONS – DESCRIPTION.

#11.6.1. MESONS ARE COLOR AND CHARGE NEUTRAL.

A meson is a color neutral combination of a quark and an anti-quark of the same color. It is made up a pair of two quarks such as $u\bar{d}$ or two pairs of four quarks such as $(u\bar{u} - d\bar{d})/\sqrt{2}$.   It is electrically neutral. Mesons have spin J of 0, 1, 2, 3, 4, 5, or 6. (Mesons with spins 5 and 6 have not yet been detected to a certainty.) See **section 10.2** for a complete listing of mesons and their quark combinations.

For example to make a meson, an up (u) quark with an electromagnetic charge of +2/3 and a color charge of red could pair with an anti-up quark with an electromagnetic charge of -2/3 and color charge of anti-red. A red colored a quark is neutralized by a red colored anti-quark.

The charm (c) and top (t) anti-quarks with an electromagnetic charge of -2/3 and a color charge of anti-red also can combine with an up (u) quark with an electromagnetic charge of +2/3 and a color charge of red to form mesons, but top quark mesons have not been detected to a certainty. (See **section 10.2**.)

#11.6.2. QUARK COMBINATIONS IN MESONS.

Each quark and anti-quark in a meson can be potentially one of six types (u d s c t b). Each meson can be of two of six types of quarks making 36 potential mesons ($u\bar{u}$ $u\bar{d}$ $u\bar{s}$ $u\bar{c}$ $u\bar{b}$; $u\bar{d}$ $d\bar{d}$ $d\bar{s}$ ... $b\bar{b}$). Each of these potential 36 mesons can each have a corresponding anti-meson making 36 potential anti- mesons ($\bar{u}u$ $\bar{d}u$ ... $\bar{b}b$). Only 19 of these different types of mesons and anti-mesons have been detected with a certainty (**section 10.2**). Two other kinds of mesons have also been detected to a certainty (**section 10.2**):

$(u\bar{u}$ - $d\bar{d}$)/$\sqrt{2}$ and $c_1(u\bar{u}$ + $d\bar{d}$) + c_2 $s\bar{s}$, where the c_1 and c_2 are c quarks

The following table summarizes the 25 mesons and the experimental detection status of them.

NOTE: In the table below, the 19 Meson types with an asterisk (*) on the right have been detected Experimentally To A Certainty" (ETAC) per the **2012 _RPP_**. All 25 meson types are listed. For clarity, only those of the 25 anti-meson types that have been detected ETAC are shown. The meson types that have been detected ETAC are listed in **section 10.2**. These 19 meson types account for the 118 mesons that have been detected ETAC and are listed in detail in the _RPP_.

u	$u\bar{u}$		$u\bar{d}$ *		$u\bar{s}$ *		$u\bar{c}$		$u\bar{b}$ *
d	$d\bar{u}$ *		$d\bar{d}$ *		$d\bar{s}$ *		$d\bar{c}$	$\bar{c}d$ *	$d\bar{b}$ *
s	$s\bar{u}$	$\bar{u}s$ *	$s\bar{d}$	$\bar{d}s$ *	$s\bar{s}$		$s\bar{c}$	$\bar{c}s$ *	$s\bar{b}$ *
c	$c\bar{u}$ * $\bar{u}c$ *		$c\bar{d}$ *		$c\bar{s}$ *		$c\bar{c}$ *	$\bar{c}c$ *	$c\bar{b}$ *
b	$b\bar{u}$	$\bar{u}b$ *	$b\bar{d}$	$\bar{d}b$ *	$b\bar{s}$	$\bar{s}b$ *	$b\bar{c}$	$\bar{c}b$ *	$b\bar{b}$

$$(u\bar{u} - d\bar{d})/\sqrt{2} \; * \qquad\qquad c_1(u\bar{u} + d\bar{d}) + c_2 \, s\bar{s} \; *$$

Physicists ordinarily refer to mesons as belonging in a meson nonet. This is a 3x3 = 9 (nonet) table consisting of u, d, and s mesons only with the c and b mesons of the 5x5 = 25 table above removed. All the mesons or their anti-mesons in this nonet have been detected ETAC as shown above. There are other ways of categorizing mesons that are also very useful.

There may be several reasons why meson type of the 25 listed above has not been found: it cannot occur, it is expensive and difficult to search for, or little or no additional scientific knowledge would be gained by searching for it. The _RPP_ is updated every two years, is due out again in 2014, and will include the latest status of theoretical and experimental particle research at that time.

#11.7. BARYONS - DESCRIPTION.

#11.7.1. BARYONS ARE COLOR NEUTRAL AND HAVE HALF-INTEGER SPIN.

Baryons are formed with color neutral combinations of three quarks. A colorless baryon has one quark of each color of three quarks (a red, a blue, an a green). Baryons carry a quantum baryon number +1; anti-baryons carry a baryon quantum number -1.

Baryons have half odd-integer spin J of 1/2, 3/2, 5/2, ... , or 15/2. (Baryons with spins of 13/2 and 15/2 have not been detected to a certainty.)

#11.7.2. PROTONS AND NEUTRONS ARE BARYONS MADE OF THREE QUARKS – WHERE DID ALL THE MASS COME FROM?

Protons and neutrons are baryons composed of quarks. Protons and neutrons along with electrons are the basis for all atoms and material in the Universe.

The proton is a positively charged (+1) uud baryon; it carries a baryon number of +1. A free proton is very stable and has a life of over 10^{31} years. (The age of the Universe is 1.38 x 10^{10} years.) A free neutron is a neutrally charged udd baryon. A free neutron is uncharged and has a life of about 15 minutes; it carries a baryon number of +1. As shown in **section 10.4** all the other baryons except the proton (LT: 10^{31} years) and neutron (LT: 15 minutes) have lifetimes (LT) of less than 10^{-8} seconds before they decay into other particles. Their exact decay products and patterns are given in the *RPP*.

An anti-baryon carries a baryon number of -1. An anti-proton carries a charge of -1.

Perhaps the reader would like to calculate the mass of a neutron or proton from the information presented so far. It seems as if we wish to do calculate the mass of a proton, all that is necessary is to add the masses of the uud quarks listed in **section 9.3**. Lets do it:

uu	= 2.3 x 2 =	4.6 MeV/c^2	
d	=	4.8 MeV/c^2	
p	= 4.6 + 4.8 =	9.4 MeV/c^2 (??)	Total mass of proton

The actual effective mass of the proton listed in **section 10.3** and determined experimentally is: 938.3 MeV/c². The "missing mass" is 928.9 MeV/c².

Where did the huge amount of extra mass (928.9 MeV/c²) come from? Most of it comes from the equivalent mass of the large amount of energy of the Strong Nuclear (Color) Force Gluons (g) that hold the red, blue, and green quarks of the proton together in a stable, color neutral configuration with a charge of +1 (+2/3, +2/3, -1/3). The force of like electric charge repulsion is the inverse of the square of the distance between charges. The closer together the quarks of like charge are, the stronger the force of repulsion.

The energy involved in holding the proton together is equivalent to mass as determined by m = E/c² (Einstein's equation. The strong force gluons provide the majority of energy and thus the energy-mass holding the proton together. Where do the gluons come from and obtain the necessary energy to hold the proton (and other particles) together? Their energy comes from the infinite reservoir of energy of empty space (**section 12.7**). There is one more source of the remaining (one percent or so) of mass of the proton: the Higgs bosons. Where do the Higgs bosons come from to give mass to the proton (and other particles)? They also come from the infinite reservoir of energy of empty space.

The strong color force bosons ensure that single quarks never exist, but are immediately joined to another quark to form a meson, or two more quarks to form a baryon such as the neutron or proton. The borrowed energy will sooner or later be paid back when the particles decay or are annihilated by their anti-particle counterparts.

A similar situation exists for the three (red, blue, and green) quarks of the color neutral neutron of charge 0 (+2/3, -1/3, -1/3). At first look, the neutron should have less mass than the proton as its repulsive charges are less and less energy might be necessary to hold it together. This is not the case. The neutron has more mass than the proton and is unstable. A neutron by itself sheds mass and decays in about 15 minutes into a proton (+1), photon, and electron (-1). An unstable particle always decays to particles of lesser energy and mass which in total have the same mass, energy, and charge.

#11.7.3. QUARK COMBINATIONS IN BARYONS.

Baryons are made of three quarks. The u, d, and s quarks combinations are:

      uuu uud udd ddd uus uds dds uss dss sss

All the above baryons have been detected ETAC.

The charge (Q) of a baryon can be -1, 0, +1, or +2; it is determined for any specific baryon by adding the charges of the quarks in the baryon. Baryons made of strange quarks also carry "S"(strange) which can be 0, -1, -2, or -3; it is determined by adding the number of s quarks in the baryon.

Additional baryons can be made of u, d, c, and b quarks and carry their numbers as well. (See **section 10.4.**) Those that have been detected ETAC are:

      udc uuc ucc ddc usc dsc ssc udb usb dsb ssb

Section 10 lists additional characteristics of baryons and other quark composites.

#11.7.4. QUARK COMBINATIONS IN MORE COMPLEX COMBINATIONS MAY BE ORIGIN OF DARK MATTER.

It is an interesting exercise to mentally invent various possible combinations of quarks that meet the rules provided in section 11.3 to produce mesons, baryons, tetra-quarks, penta-quarks, hexa-quarks, etc. Don't forget to use the top (t) quark.

The possible combinations seem endless. There are, of course, many exclusion rules that an experimental physicist would use that are beyond the scope of this book which rule out many of the combinations. But, the point is there is much work ahead for experimentalists to unlock further secrets of the building blocks of our Universe.

#12. PARTICLE INTERACTIONS – BACKGROUND INFORMATION.

#12.1. ELECTRON VOLT Ev IS USED IN PARTICLE PHYSICS FOR ENERGY = Ev, MOMENTUM = Ev/c, AND MASS = Ev/c^2.

In particle physics, the joule (J) is a very large unit of energy. The electron volt (eV) is a more convenient unit. One electron volt (eV) is the energy required to accelerate an electron through a potential difference of one volt. (See **section 1.2.15.**)

Electron charge =1 eV = 1.602 176 565 x 10^{-19} joules
= 1.602 176 565 x 10^{-19} Coulombs (C)

1 eV/c^2 = 1.782 661 845 x 10^{-36} kg

NOTE: PHYSICISTS WILL MANY TIMES OMIT THE c^2 as understood when specifying mass! (c = the speed of light.)

Electron mass = 0.510 998 928 Mev/c^2 = 9.109 382 91 x 10^{-28} grams

Proton mass   = 938.272 046 Mev/c^2 = 1.672 621777 x10^{-24} grams

 A proton weighs about 1837 times more than an electron.

K = 10^3;   M = 10^6;   G = 10^9;   T = 10^{12}

For example, the masses of an electron and a positron are about 510,999 eV. If two electrons are accelerated to high velocity and crash into each other, their charges, energy mass, and kinetic energy would have to be accounted for in their decay products (charge and energy are conserved).

The new collider at CERN is expected to produce beam energies of more than 14 TeV. Older colliders produced beam energies of just less than 1 TeV.

#12.2. CONSERVATION OF ENERGY, MOMENTUM, CHARGE, AND OTHER KEY PARTICLE CHARACTERISTICS APPLY TO EVERY INTERACTION (MEDIATION) BETWEEN PARTICLES.

Conservation of energy and momentum is a key principle of all physics. Conservation of energy and momentum apply to every interaction (mediation) of particles.

Conservation means the same after as before. It applies to the sub-atomic quantum world of elementary particles (discussed herein), the larger world of atoms composed of these particles, and the theoretical world of even tinier than particles "Strings." It applies as well to the cosmological where two planets (or galaxies) may crash into each other with some or much of the kinetic energy of the crash converted to mass according to Einstein's equation for Energy (E) equivalence to Mass (m) multiplied by the speed of light (c): $E = mc^2$. In every interaction between particles, the energy of the particles is equivalent to mass and has a gravitational attraction with other particles – in most cases, however, at the particle level this gravitational attraction is so small as to be negligible. Momentum as used herein means relativistic momentum where some or all of the mass in an interaction may be converted to energy and vice versa in accordance with Einstein's special relativity and mass energy equivalence equations.

Einstein's energy equation $E = mc^2$ shows the convertibility of mass to energy and led to the development of the atomic bomb. A hot object that takes on energy weighs more (very slightly) than a cold object. A compressed spring weighs more (very slightly) than an uncompressed spring. An electron traveling near the speed of light weighs lots more than an electron at rest.

#12.2.1. RELATIVISTIC MASS AND ENERGY OF PARTICLES ARE CONSERVED IN PARTICLE INTERACTIONS.

The relativistic equation for the mass m of an object where m_o is the rest mass of the object and v is the relative velocity of the mass with respect to the observer is:

$$m = m_o (1- v^2/c^2)^{-1/2}$$

The above equation for relativistic mass m shows that as velocity (v) increases to near the speed of light (c), total mass m approaches infinity and the energy to accelerate it to the speed of light would be infinite. That's why nothing can exceed the speed of light. This equation can be expanded using the binomial theorem:

BINOMIAL THEOREM: The binomial theorem is a method to solve equations of the general form of $(a +u)^n$. Many equations in nature including the above equation for realistic mass m are not always easy to solve exactly and result in inexact answers with many significant figures. The binomial theorem is an infinite series that allows such answers to be calculated to as many significant figures as desired. Each additional term in the series usually provides an additional significant figure of accuracy. It also lends itself to computer processing of equations that eliminates the laborious manual effort involved for complex calculations.

The first four terms in the series are:

$$(a +u)^n = a^n + na^{n-1} u + n(n-1) (a^{n-2} u^2) /1x2 + n(n-1) (n -2) (a^{n-3} u^3) /1x2x3 + ...$$

For the equation for relativistic mass $m = m_0 [1 - (v/c)^2]^{-1/2}$:

$$a= 1, \quad u = - (v/c)^2, \quad \text{and} \quad n = - 1/2$$

The first three terms for relativistic mass m are:

$$m = m_0 [1 - (v/c)^2]^{-1/2} = (m_0) [1^{-1/2} + (-1/2)(1^{-1})(-v/c)^2$$

$$+ (-1/2)(-1/2 - 1) (1^{-1/2 -2})(-v/c)^4 /1x2 + ...]$$

$$m = m_0 [1 + (1/2)(v/c)^2 + (3/8)(v/c)^4 + ...]$$

150

Substituting this into Einstein's equation for energy E:

$$E = mc^2 = m_0 c^2 [1 + (1/2)(v/c)^2 + (3/8)(v/c)^4 + \ldots]$$

This result is interesting for several reasons: The first term $(m_0 c^2)$ is simply Einstein's term for the equivalence of mass and energy. It is called the rest mass or rest energy of matter. It shows that a little bit of energy m_0 can be converted into a lot of energy as c^2 is a very, very large number.

The second term $(1/2\, m_0 v^2)$ is the non-relativistic kinetic energy of movement of rest mass, its non-relativistic kinetic energy (K).

The third and subsequent terms $[(3/8)(m_0\, v^4 c^2) + \ldots]$ are additional energies due to the relativistic effects first postulated by Einstein.

The total energy (E) of an object is the sum of an objects rest energy (E_0), non-relativistic kinetic energy (K), plus the relativistic energy-mass plus any other energy-mass of any kind of an object. An object in motion has additional mass and gravity due the equivalent mass of this kinetic energy.

This is why the equivalent mass of pure energy such as of a massless photon is attracted by other masses. Thus, light from distant galaxies and stars is bent by the Sun's gravity as it travels by on its way to the Earth. Experiments show however, the amount of bending predicted just on this basis alone is not correct and is only about half of the actual value. This is due to other effects of the energy mass of various kinds and other types of energy of the sun. Einstein's General Relativity (**section 22**) explains the actual value of the bending of light and other phenomena of gravity and acceleration as accurately incorporating all the various types of energy-mass predicted by Einstein's General Relativity equations as well as the spacetime geodesics involved.

#12.2.2. OTHER KEY CHARACTERISTICS OF PARTICLES ARE THE SAME (CONSERVED) BEFORE AND AFTER INTERACTIONS.

As described above, energy is conserved in interactions between particles and bosons. In every particle interaction, energy is conserved. That is, the total energy of the particles (kinetic, potential, and mass) before and after the interaction is the same, even though the paths of the particles may have been greatly altered. For example, the energy required to create a photon is returned when it is absorbed elsewhere and vice versa.

Several other characteristics of particles are also conserved in every interaction: including electric charge, color charge, and isospin. For example, if two electrons (negative charges) interact through the mediation of a photon, after the interaction, there must still be two negative charges with original energies conserved such as repulsion of the two electrons. Also, in an interaction mediated by the strong charge, isospin is the same before and after the interaction.

An electron (-) and a positron (+) with a net charge of zero, can interact by annihilating each other into a photon (more likely two photons) of zero charge, but having the same energy as the combined momentum and energy mass of the two particles now residing in the photons. Note that electric charge of an electron (-) and a proton (+) are exactly equal and exactly cancel. Other particles (quarks) have charges of exactly 1/3 and 2/3 the charge of an electron and proton. (See **section 9.3.**) Charge is created in exactly canceling units so the Universe and Space beyond are always exactly charge neutral.

Similarly, two photons (zero charge), with at least the minimum required energy, can interact to produce an electron (-) and a positron (+) pair of zero net charge, and the same energy as the original two photons; *thus*, creating mass out of photons (light).

Physicists describe (illustrate) the interactions between elementary particles using Feynman charts, named after their originator (**section 13.1.2**). Interactions between particles can take place in many different ways, sometimes, ten, twenty, fifty, one hundred or many, many more ways. Each way has a certain probability of occurrence. Each way has its own Feynman chart. To develop the charts and calculate the probabilities may take a very long time – days, weeks, months, or even years. When complete and done correctly, the probabilities will add up to one (1.0000) and the experimental results will agree with the calculations to an astonishing accuracy. Fortunately, after a point, inclusion of additional (less probable) ways an interaction can take place adds minuscule information, is of no value, and can be neglected.

#12.3. ELEMENTARY PARTICLES, ATOMS, MOLECULES, AND EVEN LARGE OBJECTS EXHIBIT BOTH PARTICLE AND de BROGLIE WAVE CHARACTERISTICS.

Every object regards of size displays both particle and wave characteristics. All the elementary particles and composite particles (hadrons) exhibit both particle and wave characteristics.

A particle's waves are called de Broglie waves. For most large objects, such as a planet, a moon, a mountain, a car, a rock, and a grain of sand, the wave characteristics are so small as to be imperceptible. At the atomic and elementary particle level, however, both the particle and wave characteristics of matter must be considered to understand and calculate what is going on.   For example, the particle characteristics of light photons are used to explain the photoelectric effect but wave characteristics of light photons are used to explain light refraction in a prism or eye glasses.

In 1924, Louis de Broglie suggested that matter, as well as light, might have wave properties. This would give light and matter a dual nature on equal footing. Louis de Broglie suggested a formula which was rapidly verified by experiment:

$\lambda = h/p = h/mv$ where p = momentum = mass m times velocity v, and h = Planck's constant (**section 1.2.17).**

The particle-wave characteristics apply to all quantum particles: photons, electrons, protons, neutrons, mesons, leptons, and quarks - all of the constituents of matter. They apply as well to atoms, molecules, and large objects; however, for macroscopic objects (larger than 10^{-6} meter) the de Broglie wavelengths are so small relative to the sizes of the objects as to be totally insignificant and do not show observable wave characteristics in normal circumstances.

EXAMPLE 1: The wavelength of a 1 gm mass (m) moving at velocity (v) of 1 cm/sec:

$\lambda = h/mv = (6.626 \times 10^{-27} \text{ gm-cm}^2/\text{sec})/(1 \text{ gm} \times 1 \text{ cm/sec})$
$\lambda = 6.626 \times 10^{-27} \text{ cm}$

EXAMPLE 2: For an uncharged particle of a mass about that of an electron, 9.11×10^{-28} gm moving at a velocity (v) of 1/3 the velocity of light, 1×10^{10} cm/sec:

$\lambda = h/mv = (6.626 \times 10^{-27} \text{ gm-cm}^2/\text{sec})/(9.11 \times 10^{-28} \text{ gm} \times 1 \times 10^{10} \text{ cm/sec})$
$\lambda = 60.36 \times 10^{-9} \text{ cm} = 6.036 \times 10^{-9} \text{ cm}$

12.3. ELEMENTARY PARTICLES, ATOMS, MOLECULES, AND EVEN LARGE OBJECTS EXHIBIT BOTH PARTICLE AND de BROGLIE WAVE CHARACTERISTICS (CONTINUED).

The de Broglie wavelength for an electron is 10^{18} times larger (a billion, billion times larger) than for the 1 gm mass.

The de Broglie wavelength of an uncharged particle about the size of an electron traveling at 1/3 of the velocity of light is about the same as the wavelength of an x-ray photon and has a similar double-slit interference pattern.

The double slit wave-interference pattern experiment has been carried out with uncharged particles and molecules. The result strangely is the same. The electrically neutral particle or molecule exhibits wave characteristics, seemingly passes through both slits at once, and its wave characteristics interfere with themselves to produce an interference pattern on a target screen. The interference pattern is the same as would be expected by a photon, electron, or other particle associated with electromagnetic waves.

Thus, it is not only the electromagnetic characteristics of the photon and electron that cause the interference. The quantum characteristics of electromagnetic and de Broglie wave phenomena have a common root. Both charged and uncharged particles exhibit de Broglie wave interference characteristics. Charged particles also exhibit electromagnetic wave characteristics as described by Maxwell's equations.

The de Broglie waves provide additional confirmation of the equivalence of mass and energy and the indivisibility of mass and energy beyond a certain quantum extent.

An electron with a negative (-) charge is in many instances thought of as a charged particle, that when in motion (with respect to an observer) has a magnetic field around it and when accelerated (or decelerated) radiates photons. But in an atom, an electron also exhibits wave characteristics.

As part of an atom, an electron is sometimes described as a negatively (-) charged particle orbiting the positive (+) charged nucleus like a planet orbiting a star and is subject to angular acceleration. This is not a very comprehensive description as an electron in an atom neither has a magnetic field associated with it nor radiates photons as because of its angular acceleration in orbit. A more accurate description is that the electron in an atom is a continuous vibrating energy wave that encompasses the nucleus at an allowed specific distance from it. The exact location of the electron in its orbit around the nucleus cannot be determined, and the electron has an equal probability of being found at every specific location in its orbit.

Electrons exhibit wave characteristics while in an atom. They are restricted to wavelengths (energies) that exactly fit the circumferences of the allowed orbits (allowed energy states) in an atom. Other wavelengths (energies) are not allowed and excess energy of an electron in an orbit is radiated away as photons. As an atom cools, electrons in orbit transition from one allowed energy state (orbit) to a lower energy state and radiate away a specific wavelength photon of the difference between the higher and lower energy of the allowed orbits. The radiated photons and can be detected and seen as "lines" in the radiated spectrum that are a signature of the specific radiating elements and its constituent atoms.

But here's the rub, both wave and particle characteristics cannot be investigated at the same time. Investigating characteristics of particles destroys wave characteristics and vice versa.

#12.4. UNCERTAINTY PRINCIPLE AND PROBABILITIES RULE AT THE QUANTUM LEVEL - IF SOMETHING CAN HAPPEN IT WILL, BUT MAYBE NOT VERY OFTEN.

According to the "Uncertainty Principle" of quantum physics, at the quantum (sub-atomic) level, it is not possible to know exactly one aspect about a quantum particle (such as exactly where the particle is or will go) while knowing exactly another complementary aspect about the particle such as its exact energy or momentum). Any way used to find out exactly about one aspect will change the other aspect. The best we can do is to calculate the probabilities of all the various ways that an interaction between particles or an experiment will turn out. The sum of all the probabilities must be one (1.000...). Probabilities adding up to more than one show equations of the interaction are wrong; if less than one, some possibilities or interactions have been overlooked.

#12.4.1. UNCERTAINTY PRINCIPLE.
Where exactly is a photon and where is it going? With the invention of calculus by Sir Isaac Newton and Gottfried Wilhelm Leibnitz, many physicists (including Albert Einstein) believed that if two related canonical (conjugate) variables of an object were precisely known were known at any specific time, then the exact future of the object whether a particle, ball, bullet, pendulum, auto, moon, planet, or whatever could be computed. Conditions such as temperature, windage, gravity, multiple objects, and the like could be added in as necessary. Everything that needed to be known was known - it was just a problem of measuring the objects parameters at some point of time and then doing the math. To many physicists, there were obviously no longer any mysteries about the behavior of objects.

However, as the quantum aspects were explored, the realization of that light and other objects have both and wave and particle characteristics came with a problem! It is not an insurmountable problem with large objects to measure exactly where they are at a specific time, what their velocity is, and so forth. But how do you find out where a wavelike photon or any wave is at a specific time? How do you find out its path?

12.4.1. UNCERTAINTY PRINCIPLE (CONTINUED).

Unfortunately, any way photon or particle-conjugate characteristics can be determined will require interfering with the very characteristics being determined.

For example, as described earlier (**section 12.3**), a single photon or single electron can be fired at two (or more) small slits to a screen. These particles will act as a wave and each will pass individually through all the slits at the same time, creating an interference line pattern on the screen.   Finding out which slit (or all slits at once) the photon or electron passed through cannot be done without disturbing the experimental apparatus, interfering with the path of the incident photon or electron, and destroying the interference pattern on the screen. Physicists tried many experimental approaches without success. This is not the case with large objects. Bouncing off a few photons or using another measuring technique on a large object to determine its velocity or location at specific time makes no noticeable or measurable difference in the experimental results or accuracy; or if there is an effect on the measurement, the effect can be quantified and accounted for in the results.

This uncertainty places important limitations about the precision of simultaneous measurement of the position and momentum (mass times velocity) of any object. If an object is known to be at a specific position at a specific time, then its specific momentum will be uncertain. Likewise, if its momentum is known, then its location at that specific time will be uncertain. More precisely, if one of these is known "fairly" accurately, then the other will be known much less accurately.

12.4.1. UNCERTAINTY PRINCIPLE (CONTINUED).

In 1927 Werner Karl Heisenberg expressed this situation mathematically with his Uncertainty Principle for which he received a Nobel Prize in 1932. Using a generalized notation, the Uncertainty Principle is:

$$U_{c1} \cdot U_{c2} \geq \sim \bar{h} \qquad \text{Where h is Planck's constant. } \bar{h} = h/2\pi$$

This equation means that if the measurement of conjugant variable C_1 is made with the uncertainty (accuracy) U_{c1} and simultaneously if the measurement of conjugant variable C_2 is made with the uncertainty (accuracy) U_{c2}, then the product $U_{c1} \cdot U_{c2}$ cannot be less than $\bar{h}$ or at least the order of magnitude of $\bar{h}$. Further, $\sim$ indicates that the product of the uncertainties U is at least of the order of magnitude of $\bar{h}$. Also, any one of the conjugant variables cannot be precisely measured or specified without introducing a disturbance in the measurement of its conjugant partner.

The Uncertainty Principle is also stated [**Schumm p. 70**] as:

$$U_{c1} \cdot U_{c2} \geq = h/4\pi$$

However, at Planck distances, the use of 2π or 4π is due to the mathematical derivation assumptions and is rarely significant.

Examples of conjugant variables are linear position/momentum, linear position/velocity, energy/time, angular position/velocity, angular position/angular momentum, and so forth.

Example 1: For a bullet of mass of 0.03 kg and a uncertainty U_v in its velocity measurement of 1×10^{-3} m/sec , the uncertainty U_p in its position is:

$$U_p \geq \sim \bar{h} / U_v \geq \sim 1.054 \times 10^{-27} \text{ erg sec/ } (0.03 \text{ kg})(1 \times 10^{-3} \text{ m sec}^{-1})$$

$$U_p \geq \sim 3.5 \times 10^{-30} \text{ m}$$

Example 2: For an electron of mass 0.511 MeV/c^2 (9.11×10^{-31} kg) and an uncertainty U_v in its velocity measurement of 1×10^{-3} m/sec , the uncertainty U_p in its position is:

$$U_p \geq \sim \bar{h} / U_v \geq \sim 1.054 \times 10^{-27} \text{ erg sec/ } (9.11 \times 10^{-31} \text{ kg})(1 \times 10^{-3} \text{ m sec}^{-1})$$

$$U_p \geq \sim 0.116 \text{ m}$$

For the bullet, the uncertainty in position measurement U_p relative to the position of the bullet is so small as to be beyond the capabilities of measurement equipment. For the electron, the uncertainty of position measurement is so large as to invalidate any measurement in the location of an electron and would place it anywhere among billions of atoms.

For all practical purposes, the Uncertainty Principle means at the subatomic level, the measurement process itself, of any experiment or observation, greatly affects the experiment itself. For common objects such as a bullet, the uncertainty principle does not impose any effective limits on experimental measurements; errors in position are always much smaller than one millionth of a trillionth of a trillionth (10^{-30}) meter. However, the opposite is true for objects as small as an electron.

The uncertainty principle also leads to the conclusion that in any experiment one cannot simultaneously observe the wave and particle properties of light or matter.

#12.4.2. PROBABILITY WAVES.

Particles have probabilities of being at any specific location. Quantum mechanics calculates these probabilities with an extremely high degree of accuracy. To account for these probabilities, physicists can make three dimensional drawings showing the probability amplitude of the "probability waves" at each location. The amplitude of a probability wave is proportional to the probability that the particle will be found at any specific location.

Perhaps the probability wave is a separate, distinct wave, but it seems more likely that the probability wave characteristics of a particle **are the particle**.

If we regard the electron and other (charged or uncharged) subatomic particles as waves, the results of the various experiments begin to make good sense. Regarding a subatomic particle as a wave means that even though its wavelength is very minute, the wave may extend instantly to infinity and circulate to and between all possible paths and probabilities that the particle has at once and interfere with itself under some circumstances.

This is just like aura of electromagnetic waves for the single photon created in **section 13.3** that extended to infinity. But since a probability wave does not carry energy, it is not limited to the velocity of light. It can instantly travel outwards to infinity without being limited to the velocity of light that applies to particles that an objects that consist of mass and energy. Nevertheless, when a probability wave collapses and a particle has been found to be at a specific location, its energy cannot have traveled there faster than the speed of light. Also, when a probability wave collapses, it collapses everywhere instantly. Perhaps, a way to use probability waves to communicate instantly anywhere in the Universe may be found.

An electromagnetic wave (photon) travels at the speed of light. The clock traveling with the electromagnetic wave has stopped with respect to all other clocks. The electromagnetic wave could travel to infinity while all the other clocks keep ticking. It isn't until a measurement is made of the photon's velocity, location, or other characteristic, that the photon must commit to a specific location and specific velocity (the speed of light). Then, all those other clocks will measure the photon's velocity as the speed of light even though in the interim, the photon probability wave seemingly traveled to infinity.

At the quantum level, if a particle could be found at any of 10 locations, each with some probability of the particle being at that location, the particle in fact travels to and/or is at all 10 simultaneously. Think of the particle as a probability wave. If equations and calculations are correct, the probabilities of being at all of the possible 10 locations add up to one (100 percent) or at least very close to one. It is as if the particle is a probability wave, that spreads out (travels) to all the possible locations at once and exists at all the possible (probable) locations at the same time.

It isn't until we try to find out for sure where the particle (wave) is or some event that affects the particle (wave) happens, that the particle makes up its mind and somehow selects where to be. The results of many repetitions of this event would turn out to be exactly as predicted by the probability calculations (if our calculations and assumptions are correct). Modern quantum mechanics predictions can be accurate to 15 significant decimal places.

The probability wave travels to and extends throughout the Universe (and Space beyond) to wherever there is a probability the particle may be found. The probability wave amplitude at any specific location is relative to the probability that the particle will be found at that location. The probability wave is zero at all locations where the particle cannot be found. So where exactly is the particle? The elementary particle is not only a thing or object, it is a wave. The probability wave is the particle. The particle is the probability wave. The particle is everywhere the probability wave is and on every path to all destinations where the particle has a probability to travel to and/or to be found. In some situations, it is more like a particle; in others it is a more like a wave.

No matter how far apart the ten (or however many) different locations may be, a nanometer or a light year, once the particle has chosen its final destination, all other possible and probable destinations are instantly eliminated as the probability wave collapses. This may seem to violate the axiom of Special Relativity that nothing can travel faster than the speed of light (**section 21**), as the information that a particle has chosen its destination has instantly traveled perhaps light years faster than the speed of light. However, the "probability wave" seems to be itself an energy-free, mass-free entity. It collapses instantly everywhere as an entire entity, rather than as messenger sending information faster than the speed of light. This concept carries over to the understanding of entangled particles (**section 12.6**) and particle mediation bosons including the force of gravity bosons: the Higgs boson and gravitons.

12.4.2. PROBABILITY WAVES (CONTINUED).

Our "usual" encounters with probabilities are vastly different than quantum probabilities. Usual probabilities deal with inches, centimeters, seconds, throws of dice, poker hands, and the like. Quantum probabilities deal with 10^{-33} centimeters, 10^{-40} seconds, and other very small quantities.

Our "usual" probabilities often fit a "normal distribution" where substantial deviations from the most likely expected value are often and expected. For example, throwing a coin would expect the most likely 50 percent heads and 50 percent tails to occur. Throwing dice would expect boxcars (twelve) once every 36 rolls and a seven to occur six times out of 36 rolls. In the real world, these "usual" probabilities are only expectations and rarely occur as predicted. Heads may occur 63 times out of 100 sometimes and 12 times out of 100 other times. Rarely will heads occur exactly 50 times out of 100, but occurrences near 50 times out of 100 will probably occur more often than any other outcomes. Similarly, boxcars may occur several times out of 36 throws or never occur. Rarely, a coin will land on its edge and be neither heads nor tails.

In my misspent youth as a mathematician, my wife and I were Blackjack card counters. We once saw a dealer at the Hacienda win 20 straight hands – more than a million to one probability. (Actually, I saw him win 10 straight hands (1024 to 1 against occurrence) and called my wife over from the buffet line to watch. We both saw him win the next 10 straight hands as well.) That doesn't make the dealer dishonest. Although as a mathematician, I would have convicted him in a court of law for cheating. One thing for sure, I would never play against him.

Subatomic particles are different than the objects of our usual experience. Every particle is exactly like the same particle. Objects in our usual experience have vast differences at the level of a micron (millionth of a meter). At that level, there are great differences in dice measurements, consistency of the thrower, and objects the dice encounter when they hit the table. There are skillful "mechanics" that do remarkable things with dice, cards, basketballs, or golf balls to skew the probabilities. However, they pale in comparison to elementary particles of nature that almost always throw a seven exactly six times out of thirty six and boxcars one time.

#12.4.3. WHICH OF BILLIONS AND BILLIONS OF POSSIBLE PATHS WILL A SINGLE PARTICLE SUCH AS A PHOTON TAKE?

Where exactly will a lonesome photon end up? Which of the billions of routes from its source (such as a flashlight, a dipole antenna, or a light bulb) will it take? Where is our photon located exactly in its electromagnetic energy that is traveling away from its source at the velocity of light to infinity?   If it strikes a window, will it be reflected or pass on through? If it encounters a lake will it enter the water or be reflected? If it encounters two slits, which slit will it pass through or will it go through both and interfere with itself?

Will each photon take all billions of possible paths to our detectors at once? Will each photon strike each of our billions of detectors at the same time? Or, would one photon take just one path and strike just one of the billions of detectors?

Based on the results of experiments directed at answering this question::

-Our photon travels all the possible paths at once, exists at all of them at once, and ends up at one of them only when something happens to make it select one of the billions of possibilities.

- Over many, many photons, the results will almost exactly agree with the probability predictions of quantum mechanics.

- Our photon's electromagnetic wave is or is accompanied by a "probability wave" that travels to all possible destinations at infinite velocity.

- Our photon travels and exists in all the possible paths and ends up at all of them. Over many, many photons, the results will (almost) exactly agree with the probability predictions of quantum mechanics.

- Our single photon will travel and exist in each of its possible paths at once and not make a decision as to exactly which path it will travel and be in until some outside action or intervention takes place. If only two paths were possible, then the photon travels and exists in both paths simultaneously. If an observer makes an effort to find out which path the photon took or where the photon ended up, then, and only then, will the photon actually make a choice. Until then, the photon wave circulates at each of it possible paths and exists at all its possible locations.

- One might think that it would be simple to tell which path and direction a photon was radiated. Unfortunately, it is impossible to experimentally make this determination as any experiment will require some means to perform, and this will disturb the photon under test and invalidate any experimental results.

- Some paths may be light-years longer than some of the other paths and it may be several years before a single photon traverses and arrives at all its destinations. Nevertheless, no matter how far apart the destinations are, once a choice has been made by the particle, all other destinations are instantly eliminated as destinations faster than the velocity of light!

The mathematics of quantum mechanics calculates, with remarkable and astonishing accuracy (10 significant decimal places or better), the probability that our single photon will be found at any one of its possible paths. Unfortunately, quantum mechanics does not address which path any individual photon will take or select.

#12.5. UNCERTAINTY PRINCIPLE AND dE BROGLIE WAVES EXPLAIN TWO EARLY MYSTERIES OF CHARGED PARTICLES IN ATOMIC AND HADRON STRUCTURES.

The explanation above described what is going on for "free" charged particles, but does not describe what is happening to charged particles within atoms and hadrons. The Uncertainty Principle (**section 12.4**) and de Broglie waves (**section 12.3**) provide an explanation to two early mysteries of atomic structure.

Atoms consist of charged particles, negatively charged electrons and positively-charged protons.   Mesons and baryons consist of negatively and positively charged quarks.

One electron and one proton form hydrogen, the simplest atom. Since unlike charges attract, an early mystery in development of atomic theory was why the negatively-charged electron did not radiate all its energy and fall into the positively-charged proton. This mystery was resolved by application of the Uncertainty Principle and de Broglie waves.

Electrons traveling at a constant velocity have a steady magnetic field and do not radiate energy (as photons). If the velocity of the electron varies, as it does when it is accelerated as it is in an orbit or otherwise, then electromagnetic energy will be radiated in the form of photons. So it would seem that the electrons would either fall out of orbit into the proton, or radiate energy due to the acceleration felt when orbiting the proton and then fall into the proton. But from the above, since the electron has infinite charge, it could radiate energy forever and never fall into the proton(s) by having run out of energy keeping it orbiting the nucleus.

But even if electron's infinite charge is masked and not infinite within the atom, or if the infinite positive charge of the nucleus exactly cancels out the infinite negative charge of the electron, the Uncertainty Principle prevents the electron from falling into the nucleus: As an electron gets closer to the proton, its position gets to be more and more precisely known. If it fell into the proton, its position would be (almost) exactly known. So as the electron gets closer to the proton, the uncertainly in its location gets smaller, and its momentum (p) and associated uncertainty in its kinetic energy get larger.

This increase in kinetic energy is enough to keep the electron away from the proton at least in its lowest orbit at its lowest kinetic energy level. If the electron were any closer, the Uncertainty Principle would require that the electron have a higher energy and move further away from the proton. Without this restriction, the stability of atoms and matter would not be possible in the Universe.

In addition, the electron does not radiate in orbit because the size of its orbit requires it to be considered to be a wave circulating around the proton rather than a charged particle being accelerated around a circular orbit.

Another important characteristic of electric charges within atoms is that atoms are electrically neutral as their positive charge of protons exactly cancels their negative charge of electrons. However, if two atoms are brought together, the positive charge of each atom nucleus will attract the electrons within the other atom and vice versa. This distortion of the atoms enables bonds between the atoms to form chemical compounds.

Similar considerations apply to other charged particles (particles that carry electric charge) as well. Besides the positive (+1) and negative (-1) charges of the positron and electron, various quarks carry electric charges of +1/3, -1/3, +2/3, and -2/3..

#12.6. IF STATES OF TWO OR MORE ENTANGLED PARTICLES ARE MANDATED OR PROHIBITED, ALLOWED STATES ARE INSTANTLY ENFORCED EVERYWHERE IN THE UNIVERSE WHEN ARBITRARILY SELECTED BY ENTANGLED PARTICLE.

Sometimes particles are related pairs that must have the same or opposite characteristics (values) of a characteristic such as spin or polarization. Experiments show that even if the two particles are widely separated, determining a random value of a common characteristic for one particle of the pair will immediately (faster than the velocity of light) change the value of a paired characteristic to the allowed state even if the two particles are separated by great distances. Even if one particle of a pair is "forced" to have a certain characteristic, the other particle of the pair no matter how far away is always found to instantly have the allowed value of the characteristic.

Similarly, sometimes several related (entangled) particles can only exist with the same or allowed values of various characteristics. If one related particle has red "color," two other related particles may be allowed to have only different values of the characteristic such as blue and green color. Experiments show that even if the related particles are widely separated, determining the characteristic for one particle of the related particles will immediately (faster than the velocity of light) change the same values of characteristic of the other related particles to conform to the allowed values even if the three related particles are separated by great distances. Even if one particle is "forced" to have a certain value of a characteristic, the other particles of the pair are always found to instantly have the allowed conforming values of the characteristic.

It's as if you tossed a coin up and it randomly came up heads, or tails, or stood on edge, a coin randomly tossed up somewhere else light years away, instantly (faster than the speed of light) always came up exactly the same as the coin you tossed up.

For entangled particles, when one of the entangled particles arbitrarily selects a state, it instantly forces all associated entangled particles everywhere to take on their allowed states and prevents them from taking on prohibited states no matter how far the particles are apart. This occurs instantly. This entanglement comes about in every interaction of particles by virtue of exclusion principles and conservation of various parameters of particles in an interaction.

Similar to the seemingly arbitrary selection of a probable location by a particle instantly eliminating all other probable locations (**section 12.6**), no matter how far apart the different locations of the entangled particles may be, a nanometer or a light year, once an entangled particle has chosen its arbitrary state, all other entangled particles instantly fall into line, take on their allowed states, and eliminate their prohibited states as the entanglement wave collapses. As with probability waves that may seem to violate the axiom of Special Relativity that nothing can travel faster than the speed of light, the information that a particle has chosen its arbitrary state has instantly traveled perhaps light years faster than the speed of light,. However, the "entanglement wave" like the "probability wave" seems to be itself an entity. It collapses instantly everywhere as an entire entity, rather than as messenger sending information faster than the speed of light. Thus the speed of light is not violated and nothing is "sent" faster than the speed of light.

A particle must, therefore be thought of not only as a particle but in addition as a combination of several wave characteristics: de Broglie wave (**section 12.3**), probability wave (**section 12.4.2**), and entanglement wave (this **section**).

#12.7. EMPTY SPACE IS NOT EMPTY.

Before our Universe, what existed? One hypothesis, described in this book, is that infinite empty Spacetime always existed, and will always continue to exist. (The term Spacetime is used to include the effects of General and Special Relativity (**sections 21 and 22**). Our Universe was created and its Big Bang took place in a tiny little sphere within infinite Spacetime. Our Universe is now unfurling into infinite Spacetime at an accelerating rate.

And if we jumped in our hyper- speed spacecraft and traveled faster than the speed of light to the frontiers of our expanding universe, we would find "outer spacetime" beyond our Universe to be just about like "empty spacetime" is in our Universe, less all the various galaxies. Beyond the frontiers of our Universe, our clocks would still keep ticking just as they do in empty spacetime in our Universe. Beyond our Universe there are perhaps other universes, maybe an infinite number of other universes. Temperature in empty space would be much, much less than 1 K, almost 0 K (10^{-30} K?).

Another viewpoint (perhaps more mainstream) is that nothing exists beyond our Universe (EXCEPT PERHAPS MORE UNIVERSES). Time and Spacetime do not exist until our Universe unfurled at the Big Bang and continues to unfurl to create it. Perhaps a problem in this viewpoint is that before the Big Bang there was something - something that the big Bang was conceived in and born in. Which viewpoint is correct may not be determined during our lifetimes and may never be determined.

Some physicists have calculated that there are potentially 10^{500} different kinds of universes possible each with a different set of laws of physics. It would seem that there should be a different kind of empty spacetime for each kind of universe. So far none of these conjectures have been blessed with any kind of confirmation.

#12.7.1. EMPTY SPACE WITHIN OUR UNIVERSE IS NOT EMPTY.

Nevertheless, empty "spacetime" within our Universe is not empty. It is teaming with virtual particle and anti-particle pairs that come into being by "borrowing" energy from empty Spacetime.

Empty spacetime is an infinite source of both positive and negative energy (**section 2.10.3**). Positive or negative energy may be borrowed but it must always be associated elsewhere with a deficit of an exact same amount of negative or positive energy. Overall spacetime is exactly energy neutral. During the inflation of the Big Bang (**section 6.2), the negative energy of our inflating Universe was borrowed and converted to positive energy and mass of our Universe**

The virtual particles and anti-particles of borrowed energy usually exist for a short time, and then annihilate each other into a puff of cancelling positive and negative energy - photons or other particles, anti-particles and bosons that ultimately are absorbed or annihilate each other. The puffs of energy are just the right amount to exactly the repay the borrowed positive and negative energy.

Measuring zero energy everywhere in empty spacetime would violate the Uncertainty Principle (**section 12.4.1**), as then the exact energy everywhere would always be exactly known, so some background amount of positive and negative energy must be measured in any small region of Spacetime, but overall the net energy of Spacetime is zero.

Empty spacetime in our Universe is an infinite reservoir of energy. It is an infinite reservoir of both Positive and Negative energy (**section 2.10.3**). The deal is that energy can be borrowed at any time in any amount but it has to be paid back sooner or later. Take a cupful of positive energy out of empty spacetime and it leaves behind a cupful of negative energy. Empty spacetime demands a borrowed cupful be paid back sooner or later. Sooner or later those virtual particles will annihilate or decay and sooner or later will give back their energy to the great reservoir of empty spacetime. Sooner or later, spacetime must get its due.

Free quarks are not found in nature. Strong nuclear force "virtual "gluons of empty space will find "free" single quarks "instantly." Then gluons will equally quickly find one or more other quarks to bond with from the turmoil of virtual particles in empty spacetime. Free virtual quarks will instantly be annihilated by a virtual anti-quark or combine with a single quark to form a meson or two quarks to from a baryon such as a proton.

12.7.1. EMPTY SPACE WITHIN OUR UNIVERSE IS NOT EMPTY (CONTINUED).

The seething energy of empty Spacetime within our Universe is real!! It can be seen in various ways:

(1) The electrical charges in empty Spacetime mask the true infinite charge of the electron and quarks **(section 13.3.2).**

(2) The Casimir effect displays the force of the electrical charges of empty Spacetime. In the Casimir effect, two electrically neutral, perfectly flat, highly conductive metal plates are slowly moved together. Every physicist and engineer knows that the electric field at the surface of a thin conducting metal plate is zero as it is cancelled by the free electrons in the metal. Only wavelengths that exactly fit between the plates and are zero at the plates can exist between the plates. This is analogous to a vibrating rod, that when struck by a hammer vibrates at a frequency determined by its length; only wavelengths or sub-multiples of the wavelengths that exactly fit on the rod can exist on the vibrating rod.

As the metal plates are moved together, only wave lengths that are zero at the plates can exist within the plates and can resist the electric fields outside the two plates. The force of the unrestricted electric fields outside the plates then pushes the plates toward each other. This force, the Casimir Effect, can be measured showing the virtual charges and electric fields within empty spacetime.

(3) An interesting sidelight to the Casimir effect is that the free electrons in the conductor that cancel any electromagnetic field at the boundary of the metal plate and spacetime, also screen the photons in the conductor, prevent their electromagnetic field, and give the photons (which are massless) an "effective" mass. This effective mass of photons is used to carry out various calculations about photons in a conductor. This "screening "effect also gave physicists insight that somehow screening electric charge and gravitational charge could resolve problems in understanding how infinite electric charge is screened to its measurable effective charge and thus how mass charge might occur in nature and led to theorizing the Higgs boson. (See below and **section 13.7.**)

(4) Virtual particle pairs, besides popping in and out of existence in "not-so-empty" empty spacetime, pop in and out of existence within an atom. Virtual electron and anti-electron (positron) pairs may pop in to and out of existence. A virtual electron is closer to the protons in the nucleus, the virtual positron closer to the orbiting electrons. If the atom is heated, the atom takes on energy and its electrons move to higher energy orbits. As it cools, electrons move to lower energy orbits and the atom emits energy by giving off photons (light) at specific frequencies that make up the atom's spectrum.

The virtual electrons and positrons distort the spectrum, giving the atom what is called the fine structure of the spectrum of the atom. This fine structure is conclusive experimental proof of the existence of virtual particles.

The only difference between particles and virtual particles is that virtual particles are usually short lived and regular particles live longer. "Short" and "long" are relative. For example, an electron e$^-$ and positron e$^+$ pair may pop into existence and annihilate each other almost immediately. Or the pair may pop into existence in "empty spacetime," become separated, and enjoy a much longer life - perhaps an infinite life.

(5) The Higgs bosons and related fields are important in empty spacetime. As explained in **section 13.7**, the Higgs boson was theorized as part of a search to find how the weak force W$^-$ W$^+$ Z^0 bosons obtained mass - no other bosons (photons and gluons) had any mass. The Standard Model (SM) of particle physics was seen as deficient and of questionable validity until the existence of four Higgs fields was theorized to give mass to the weak force bosons and other particles along with the existence of Higgs bosons. The Higgs fields are separate from the fabric of empty spacetime that is described in this **section** and elsewhere, but are just more characteristics of the fabric of empty spacetime of our Universe and Spacetime beyond our Universe.

#12.7.2. EMPTY SPACE BEYOND OUR UNIVERSE (IF IT EXISTS) IS NOT EMPTY.

The fabric of empty spacetime beyond our Universe (if it exists) would seem to be just like the fabric of our "unfurling" Universe described above **in section 12.7.1**. When future generations travel beyond our Universe, they may find empty spacetime there is just like empty spacetime of our Universe – except perhaps much colder (**section 1.2.23**). Empty space beyond our Universe would be a source of infinite negative energy (**section 2.10.3**). The laws of physics may be (or may not be) the same and mass may (or may not) be a characteristic of the fabric of spacetime there like it is in the fabric of Spacetime in our Universe. See **section 13.7**.

Earlier (**section 6.5**), the Higgs fields in empty spacetime are described as possibly undergoing a change-in-state at temperatures near absolute zero (10^{-30} K), so that they did not function to impede changes in velocity and acceleration. At almost at absolute zero temperature, the Higgs fields may be "frozen and slick like ice" and allow our Universe to inflate faster than the velocity of light at the earliest moments of its birth.

#13.   PARTICLE INTERACTIONS: THE HEART OF PARTICLE PHYSICS.

From the previous **section**s, all the pieces are now in place to accomplish the objective of placing all the elementary particles of physics including the Higgs bosons and gravitons in their proper place in particle physics. This task is simplified because with some yet unproven exceptions **(section 14),** everything in the Universe and empty Space beyond is made of these particles.   As was pointed out earlier **in section 8**, besides the electron and three neutrinos, out of the 283 "confirmed ETAC" hadrons that our Universe is made up only two are around more than a fraction of a microsecond: the proton (seemingly infinite live time) and neutron (14.67 minutes unless it is confined within an atom). The proton and neutron are in turn are made up of quarks. The rest of the hadrons (**section 10**) aren't around very long as their lifetimes are only a minuscule part of a second. Add some electrons to protons and neutrons, and you can make all the atoms that are in the Universe. An electron is an electromagnetically charged lepton. Like the proton, it is long-lived (infinite), but the other two electromagnetically charged leptons, the *Muon* and Tau, decay in much less than a fraction of a second

This **section** describes how elementary particles interact with each other, how the elementary particles interact to combine to produce hadrons, how hadrons decay into stable particles, and how the Higgs bosons and gravitons impart mass to particles.

#13.1. WAYS THAT PARTICLES INTERACT.

When two particles approach each other, an interaction takes place. Particles can approach each other and then scatter (deflect away from or toward each other); particles can bind together and create hadrons; and particles particularly hadrons can decay into other particles.   At the submicroscopic level of particle physics, conservation rules are the only way to tell exactly what goes on in any interaction between particles. Experimental physicists precisely measure all the characteristics of particles that are formed in particle interactions, usually when streams of extremely high-velocity electrons or streams of extremely high-velocity protons crash into each other or targets. Conserved characteristics of energy, momentum, charge, spin, and other parameters that are listed in the following **section**s would be carefully measured.

Particle interactions are initiated in various ways. The latest method is two ultra-high-energy streams of protons are crashed together at the particle collider at CERN. Cosmic rays bombarding earth are also a readily available source of sporadic high energy electrons, protons, and other nuclei and particles at high altitude laboratories such as at the top-floor Physics Laboratory at the University of Colorado in Boulder which I used for a short time many years ago. After interaction(s), characteristics of the all particles that are left over and any new particles that are produced are precisely measured. Anomalies in the expected outcomes of interactions have in many cases resulted in discovery of new particles.

In some cases, the anomalies were due to errors in calculations or experimental technique. In many cases, theoreticians have predicted an outcome as the energy of the collider beams has been increased. This was the case with observation of the predicted Higgs boson at CERN in 2012. Observing the Higgs was a great victory for and confirmation of the SM. Had the Higgs not been found, the SM would have been proven to be incorrect and dozens and dozens of years of work by thousands and thousands of physicists would have had to be rethought and redone.

For, example, electric charge (q) is a particle characteristic that must be conserved in interactions between particles. If high-energy streams of protons (with charge of +1) are crashed together, their three quarks with charges of +2/3, +2/3, and -1/3 interacting through strong-nuclear color force bosons, will always produce products with the have a total charge of +1 whether various other particle are mesons, baryons, electrons, positrons, photons, or whatever.

#13.1.1. INTERACTION NOTATION.

The following notation is used in equations to describe the interaction of particles. Note that charge and color are conserved in each interaction below. Preparation of these equations must make sure that all conserved characteristics as described in the remainder of **section 13** are properly accounted for on both sides of an equation.

Charged electron and a charged positron can interact through a photon(s) and produce two uncharged photons. Not that the + and − electric charges cancel each other.
$$e^- + e^+ + \gamma \rightarrow \gamma + \gamma$$

Two photons can produce an electron and a positron: $\gamma + \gamma \rightarrow e^- + e^+$
creating matter (electron and positron) out of energy (photons).

Matter and anti-matter can change matter into energy (photons): $e^- + e^+ + \gamma \rightarrow \gamma + \gamma$

A photon interacts with an electron to scatter it (change its course) in a direction depending on the relative momentum of the two particles. The electron now carries all the momentum: $e^- + \gamma \rightarrow e^-$

Two up quarks u (charge +2/3 each) and one down quark d (charge -1/3) interact through a gluon g to produce a proton p(uud) (charge (+1):

$$u(+2/3) + u(+2/3) + d(-1/3) + g \rightarrow p(uud)(+1)$$

Or: more precisely, three quarks, each of a different color interact through a gluon to produce a colorless proton:

$$u^r + u^b + d^g + g \rightarrow pg$$ The gluon(s) remain within the proton to bind the three quarks together.

A neutron (udd) decays (mediated by a Z^0 boson) and changes to a proton (uud), electron, and anti-electron-neutrino. The proton weights less than the neutron.

$$n + Z^0 \rightarrow p^+ + e^- + \bar{\nu}_e$$

An electron and positron (mediated by a photon) create two quarks q (if the electron and positron energy is high enough) which then quickly (with binding of a gluon) combine to form a $q\bar{q}$ meson:

$$e^- + e^+ + \gamma \rightarrow q + \bar{q} + g \rightarrow q\bar{q}\, g$$

Showing the bosons (g , Z^0, γ) such as in the last four examples is included here for clarity. This level of detail is usually left to Feynman diagrams which are described in the following **section**. The W^+ W^- Z^0 weak force bosons are involved in some cases when necessary to maintain the conservation of charge from one side of an equation to the other as they can bring charge to an interaction or absorb charge to carry charge away.

The weak charge bosons then decay into other appropriately charged particles. Both charged weak force bosons (W^+ and W^-) being involved in a interaction at the same time is indicated as W meaning both W^+ and W^-. If the charge of a quark is not involved or does not change from one side of the equation to the other, then the Z^0 boson will mediate the interaction.

From **section 10.1**:

π^+ Quarks: $u\bar{d}$ Q: 1   SCB: 0 0 0   M: 0139.57018   I: 1   J: 0
LT: 2.6033×10^{-8} Decay: $\mu^+ + v_u$

The π^+ meson consists of $u\bar{d}$ quarks (with a total charge Q of +1 (+2/3 and +1/3), with a LT of 2.6033×10^{-8} seconds. Per the **RPP**, the π^+ meson decays into a anti-Muon μ^+ and Muon neutrino v_u (mediated by W bosons):

$$\pi^+ + W \rightarrow \mu^+ + v_u$$

13.1.1. INTERACTION NOTATION (CONTINUED).

From **section 10.3:**

BARYON: N   **_QTY:_ 17** **_SYMBOL:_** p, N$^+$, n, N^0 **_QUARKS:_** uud, udd
S = 0 **LT:** 880.1 and 10^{32} I = 1/2   J = 1/2, 3/2, 5/2, 7/2, 9/2, 11/2

As shown above, the **_RPP_** lists a **QUANTITY** of 17 different kinds of **N** baryons. The neutron **n** is the most familiar **N** baryon.

The **_RPP_** lists the n (neutron) Decay Modes as: p$^+$ e$^-$ $\bar{v}_e$ 100 %.
A sub-listing indicates that a photon may also be present: p$^+$ e$^-$ $\bar{v}_e$ Υ.

A neutron n (udd) decays (mediated by a Z^0 boson); the neutron decays into a proton (uud), electron, an anti-electron neutrino and sometimes a photon or two to carry any excess energy away.

$$n + Z^0 \rightarrow p^+ + e^- + \bar{v}_e + Υ + Υ$$

#13.1.2. FEYNMAN DIAGRAMS

Physicists use a more complicated type of diagram to show interactions called Feynman Diagrams. An interaction may take place in hundreds of different ways. See the *RPP* for complete information on all mesons and hadrons. Long before the experiments shown in the *RPP* are carried out, physicists calculate the probably of each type of interaction, the particles produced, and the characteristics conserved; this may have taken years to calculate in the case of complex interactions.

#13.2. PARTICLE AND ANTI-PARTICLE INTERACTIONS.

Every particle has its anti-particle. A particle and anti-particle may have opposite electric charges or be uncharged. Particles and anti-particles quickly annihilate each other in some manner when they come in contact. They may decay and/or produce pure energy in a form which may then decay further. An anti-particle is usually indicated by a bar over the symbol for the particle. There are a few exceptions to this rule given in **section 11.1.**

Some particles (photons, neutrinos, and gluons, and weak force boson) act as their own antiparticles. Some particles, for example, a photon and its anti-particle another photon very seldom annihilate each other because they are not coherent. If they did, there would be no light and we would be unable to see. However, under some conditions, two photons can annihilate each other and produce an electron and a positron pair (matter created out of light). Under other conditions, two photons can interfere with each other and produce interference patterns by cancelling and reinforcing their electromagnetic waves. In some cases, a photon and its anti-photon can exactly cancel each other and disappear into nothingness.

Each boson (**section 11.4**) could also be partially or exactly canceled or reinforced under the right conditions depending on relative amplitude, phase, and wavelength of another boson just as two ocean waves cancel or reinforce each other depending on their relative amplitudes, phases, and wavelengths.

The neutron is typical of a type of particle which does not have electric charge, yet has an anti-particle. The anti-particle in this case carries other quantum characteristics such as baryon number that change sign for anti-particles. A neutron, even though it is neutral, consists of one u quark with charge of +2/3 and two dd quarks with charge of - 1/3 each so it is electrically neutral. An anti-neutron consists of one $\bar{u}$ anti-quark with a charge of -2/3 and two $\bar{d}\bar{d}$ anti-quarks with a charge of +1/3 each so it is electrically neutral also. One d quark is at the center of the neutron, the other d quark is at the surface of the neutron, and the u quark is in between.

Just as a proton can combine with an electron to form the atom hydrogen, an anti-proton can combine with an anti-electron (positron) to form anti-hydrogen. This anti-atom and others rarely if ever occur naturally but have been made in the laboratory. More complex anti-atoms of anti-protons, anti-neutrons, and anti-electrons (positrons) can also exist and have been manufactured in the laboratory.

For some reason, perhaps by accident or otherwise, our Universe is literally entirely populated by matter. Anti-matter is rare and when it exists, it is quickly annihilated by matter. It seems likely that equal amounts of matter and anti-matter were produced during the Big Bang (**section 6**) but as the Universe expanded by inflation, more matter was created than anti-matter because of the unsymmetrical decay of K baryons (kaons) resulting in a preference for matter over anti-matter. Theoretical analysis and experimental results seem to bear this out.

Whether this is accidental or natural, is not known or understood. If natural, universes of anti-matter may not exist. It seems likely that this is the case.

An anti-particle will annihilate its particle counterpart converting it into an equivalent amount of energy, if, and only if, it contacts its counterpart particle under the right conditions. And vice versa, a particle will annihilate its anti-particle counterpart if, and only if, it contacts its counterpart anti-particle under the right conditions.

When particles and anti-particles come together and annihilate each other, their combined mass is converted into an equivalent amount of energy, in accordance with Einstein's equation, $E = mc^2$. This energy then goes on to form of other particles and bosons according to the allowable quantum combinations and probabilities until stable configurations are reached with the mass, energy, and momentum of the final products the same as the original particles.

#13.3. ELECTROMAGNETIC FORCE BOSONS (PHOTONS) MEDIATE INTERACTIONS BETWEEN CHARGED PARTICLES - COULOMB'S LAW.

Familiar visible light is electromagnetic waves of photons. Without photons, we would be unable to see anything at all.   Radio and TV waves are also photons. So are x-rays. So are microwave-oven microwave waves. So are gamma rays. The frequency spectrum of photons is an almost infinitely wide range (**section 1.2.12**). It ranges from much less than a few Hertz (cycles per second) such as 60×10^0 Hz (60 cycles-per-second) electrical power in homes to more than 10^{21} Hz gamma rays. Only a minuscule part of the possible spectrum of photons is visible, yet when the term "light photon" is used by physicists, it may loosely mean the complete spectrum of visible and not visible photons.   The energy of a photon is proportional to it frequency **v** and inversely proportional to its wavelength **λ**:

$$\lambda = c/v \qquad \text{where } c = \text{velocity of light (299.792} \times 10^6 \text{ meters per second).}$$

The energy E of a photon is proportional to its frequency:

$$E = h\,v = hc/\lambda \qquad \text{where h is Planck's constant. } h = 6.626 \times 10^{-27} \text{ erg-sec}$$

A cosmic ray photon ($v = 10^{21}$ Hertz = 10^{21} /sec) has "enormous" energy E_{cr} :

$$E_{cr} = (6.626 \times 10^{-27} \text{ erg-sec}) \times 10^{21} \text{/sec} = 10^{-6} \text{ erg}$$

A visible light photon ($v = 10^{14}$ Hertz = 10^{14} /sec), has the energy E_{vl} :

$$E_{vl} = (6.626 \times 10^{-27} \text{ erg-sec}) \times 10^{14} \text{/sec} = 10^{-13} \text{ erg}$$

The photon is the most observable elementary force particle. We see it everywhere as light, electricity, and magnetism. In many situations, the electromagnetic force is stronger than the force of gravity. Photons also act as a mediator force between particles with electrical charge like the electron and proton. Remember, the electron while being a charged particle, also behaves as if it is either a particle or a wave of energy that reacts to photon as if it is a particle or wave as if it is a wave as well.

 Like electric charges, such as between two proton waves (+, +) or between two electron waves (-, -), repel (are pushed apart). Unlike charge waves such as between a proton (+) and electron (-) attract (are pushed together). Like charges add to each other. Unlike charges cancel making atoms and large masses electrically neutral.

COULOMB'S LAW – FORCE BETWEEN ELECTRICALLY CHARGED PARTICLES.

The charge (Q) of an electrically charged particle is quantized and polarized either positively (+) or negatively (-). (See **section 9.2**.)   For example, electrons carry negative (-) charge and protons carry positive (+) charge. There is a force **F** between charged particles. The force attracts (pulls the two particles together) if the particles have opposite polarity charges. The force repulses (pushes the two particles apart) if the particles have the same polarity charges.

The magnitude and direction of the force **F** is dependent on the amount of charge (Q) of each of the two particles Q_1 and Q_2. Similar to the Force of gravity (**section 2.3.3**), electric Force **F** varies inversely with the distance between the centers of the two charges.

Coulomb's Law expresses the force between two charged particles: $F = k\,(Q_1\,Q_2)/r^2$

Charge Q of electron or proton$= 1.60217657 \times 10^{-19}$ Coulombs $= 1.6 \times 10^{-19}$ C
$k = (2.9917925)^2 \times 10^{9}$ Newton meter^2 /C$^2 = 9 \times 10^{9}$ Nm^2 /C^2
For an electron (Q_1) and proton (Q_2) 1 meter (m) apart:
$\quad$ $F = k(Q_1\,Q_2)/r^2 = (9 \times 10^{9}$ Nm^2 /C^2) (-1.6 $\times 10^{-19}$ C)(1.6 $\times 10^{-19}$ C)/1^2 m^2
$\quad$ $= -23.04 \times 10^{-29}$ N $= 23.04 \times 10^{-29}$ Newton

See **sections 13.3.1 through 13.3.3** for a detailed description of how charged particles repel or attract each other. It's worthwhile to elaborate there for three reasons: First, the discussion sets the stage for understanding the rest of the forces and bosons and, in particular, the Higgs boson. Second, it provides a detailed **physical** description of what is going on in particle interactions.

#13.3.1. ELECTROMAGNETIC CHARGES OF ELEMENTARY PARTICLES ARE INFINITE TO ENABLE ELECTROMAGNETIC INTERACTION.

So what happens when a "free" electron (-) or proton (+), or other charged particle minding its own business floats around through space without another electron (or proton) or other charged particle anywhere around? The usual discussions state something like throwing balls (photons) back and forth: "Two electrons approach each other.   The electrons throw out (emit) virtual photons (bosons) that causing them to recoil (conservation of energy). The process is continued with more photons being thrown out and the two electrons then veer away from (are repelled by) each other."

The process is actually more complicated. The above explanation requires each electron to have some sort of unspecified radar that detects the other electron and tells it where to send out photons to repel the other electron. These were not problems in pre-quantum physics where each charged particle had an infinite positive (+q) or negative (-q) aura or electric "field" (E) all around it which had a field intensity that decreased in accordance with the inverse square of the distance (r) away from the charged particle: $E = q/r^2$.

That explanation was fine in pre-quantum physics days, but it is not satisfactory today. It just doesn't get to the heart of the matter.

To start with, where did this poor lonely electron charge and probability wave, just minding its own business, come from, where is it going, and where does it want to go?

As explained in **section 12.7,** empty space is not empty. It is an infinite source of positive and negative energy. It is seething with virtual particles (quarks, electrons, positrons, etc.) that pop in and out of existence. The energy to create these particles comes from the infinite energy of empty space.

Nevertheless, empty space is energy neutral; the positive and electric charge when summed over all empty space is zero. So the negative charge energy to create this lonely electron has been borrowed from the energy well of empty space and has to be returned to empty space eventually. Otherwise, there would be a huge energy imbalance or electric charge build up in empty space. In the infinite years empty space has been around that hasn't happened!

So this electron is determined to hook up with a positron, proton, or other positive charge to neutralize its negative charge with a positive electric charge or to annihilate itself and pay back all its energy and electric charge to infinite empty space. To do this, it sends out a search party in the form of an electric field aura to find a positive charge.

The electric field and electromagnetic waves are the electromagnetic boson quanta: photons. It takes energy to create photons. They have energy even though they have no mass. It takes a lot of energy to create an aura of an electromagnetic-wave field of photons all around an electrically charged particle that radiates out in all directions at the speed of light **to infinity**. This aura of photons wants to find a second electrically charged particle to interact with it, absorb a photon, and send out a photon so as to interact with the first charged particle.

How can this be you might ask, "Wouldn't this take an infinite amount of energy? Wouldn't the charge of an electron have to be infinite?" The answer is yes to both questions. The electrical charge of a charged particle is infinite. Remember however, the conservation of charge. Positive charge is always created with an exactly equal negative charge, the net charge being zero. The infinite energy to create electromagnetic fields are photons the electron sends out using energy "borrowed" from the infinite energy of empty space. (An infinite amount of energy can be borrowed from a source of infinite energy an infinite number of times.)

#13.3.2. PROPERTIES OF EMPTY SPACE SCREEN TRUE VALUE OF PARTICLE ELECTRIC CHARGE.

The properties of space, the teaming activity of virtual charged particle pairs in empty space (**section 12.7**), and the radiated aura of electromagnetic photon waves screen the true value of infinity of an electric charge so only the usual (effective) charge of an electron, quark, proton or other charged particle is measured and measurable.

To understand why the infinite charge of an electron, or other charged particle is infinite, think of a negative charged electron. All the virtual positive charges around are attracted to it and form a positively charged halo around it. This halo reduces the charge that can be measured to its effective charge, the value we can actually measure.

Experiments have shown that as extreme conditions are set up, the measured electric charge increases and at the extreme may approach infinity. But, not to worry! For every infinite positive charge there is an exactly the same infinite negative charge out there somewhere and vice versa, so the Universe and Space beyond will always have a net electric charge of zero.

#13.3.3. PHOTON SPIN ENABLES ELECTRIC CHARGES TO ATTRACT OR REPEL EACH OTHER.

Photon waves of our electron's electromagnetic field are created with "borrowed" energy from the electron's infinite charge to send out the aura of photon waves from the charged particle to infinity returning some of the "borrowed energy" to infinite empty space. The particle it is interacting with also sends out electromagnetic waves that pay back energy. Then the photon waves are absorbed by the two charged particles involved.

So here's the problem now. Our electron has sent out an aura of electromagnetic field of photons in all directions. Some have been received by another charged particle that is also sending out an aura of electromagnetic field of photons to our electron that receives (intercepts) them. But how do these two particles know whether to be attracted to each other or repelled???

It's like dating on the internet. Who for certain is on the other end? For internet dating perhaps Skype_{TM} solves the problem. So also, our electron must know what is on the other end of the electromagnetic field it received so as to be "attracted" to a positive charge, or be repelled by a negative charge.

To determine how this comes about. look at the characteristics of the electromagnetic photon in **section. 9.4.1.** The spin of the photon is listed as 1. This could mean +1 (clockwise), 0, or -1 (counterclockwise). However, for the photon, only two spins are possible, +1 (clockwise) and -1 (counterclockwise). Some particles can have the third possibility, but not a photon. Polarization is also a possibility, but vertical or horizontal polarization has no meaning in space as space has no up or down or right or left, so J spin or circular polarization is the likely solution.

Thus, the **spin of a photon (clockwise or counterclockwise) is the carrier of the polarity of the source of the photon aura emanating from a charged particle**.

If the **receiver** of our electron's counterclockwise (**negative**) aura photon is a **positively** charged proton, it will absorb the photon and "**kick out a virtual photon in a direction away**" from our electron. The **virtual photon causes the proton to move toward our electron** seeming **attracted toward it**. (If you shoot a rifle at a target in front of you, conservation of momentum causes the rifle to "kick" and drive you in the opposite direction.) Likewise **the particle expelling a virtual photon will always move directly opposite to the direction of the expelled virtual photon.**.

If the **receiver** of our electron's counterclockwise (**negative**) aura photon is a **negatively** charged electron, it will absorb the photon and "**kick out a virtual photon in a direction toward**" our electron. The **virtual photon, in this case causes the electron to move away from our electron** seeming **repelled by it.**

Electromagnetic photons are radiated out between two charged particles, are absorbed, and a virtual photon is emitted (kicked out) emitting virtual photons in a direction to either attract unlike charges (pushing the particles together) or repelling like charges (pushing the particles apart). Momentum and energy in the interactions are conserved to push the particles apart or pull them together. The total net energy and linear momentum of the two particles in the interaction is the same before and after the interaction, just as it is if one billiard ball strikes another.

#13.4. STRONG NUCLEAR (COLOR) FORCE BOSONS (GLUONS) HOLD QUARKS TOGETHER.

The strong nuclear color force holds quarks together that make up mesons, protons, neutrons, and other baryons. The strong nuclear force prevails despite the repulsive electromagnetic force bosons of the positive electric charge of quarks trying to tear the quarks apart. Leptons do not carry color charge and do not participate in strong nuclear force interactions.

The strong force is long range, but usually operates at short range. It is effective for about 10^{-15} meter (10^{-17} centimeter) - about the diameter of interior of a hadron such as a neutron or proton. The strong force never gets to act at a long range as free quarks just don't have an opportunity to exist.

#13.4.1. STRONG NUCLEAR (COLOR) FORCE CONSISTS OF EIGHT GLUONS.

The strong nuclear force is also called the color force. It holds together mesons that are made of two quarks and hadrons such as protons and neutrons that are made of three quarks. All quarks are carriers of the strong force (color charge) just as the electron is carrier of the electromagnetic force (electromagnetic charge). Some quarks are also the carrier of the electromagnetic force. (See **section 9.3.**) The colors of the color force are red, blue, and green (r, b, g).

Each quark carries one color or one anti-color. A color is cancelled by an anti-color. There are nine combinations of these three colors taken two at a time (r $\bar{r}$, r $\bar{b}$, ...) Unfortunately, this is **not the way the color force works. The strong nuclear color force is made up of a color octet of eight color bosons** [Griffiths, p. 285] **as shown below:**

$$(r\bar{b} + b\bar{r}) \sqrt{2} \qquad -i\,(r\bar{g} + g\bar{r}) \sqrt{2}$$

$$-i\,(r\bar{b} + b\bar{r}) \sqrt{2} \qquad (b\bar{g} + g\bar{b}) \sqrt{2}$$

$$(r\bar{r} + b\bar{b}) \sqrt{2} \qquad -i\,(b\bar{g} - g\bar{b}) \sqrt{2}$$

$$(r\bar{g} + g\bar{r}) \sqrt{2} \qquad (r\bar{r} + b\bar{b} - 2g\bar{g}) \sqrt{6}$$

This color octet of gluons is an outcome of the mathematics of Quantum Chromo-Dynamics (QCD) as an SU(3) color octet which has been verified by the experimental results in **sections 9.3 and 9.4.3.**

[Baggott 2, p. 143] shows the eight color bosons more simply as:

r $\bar{b}$ r $\bar{g}$ b $\bar{r}$ b $\bar{g}$ g $\bar{r}$ g $\bar{b}$ d1   d2

Baggott's approach ostensibly combines two color bosons into a color and anti color to provide three that are missing from the above list:

$$r\bar{b} + b\bar{r} \rightarrow r\bar{r} + b\bar{b} \qquad\qquad b\bar{r} + r\bar{b} \rightarrow b\bar{b} + r\bar{r}$$
$$g\bar{r} + r\bar{g} \rightarrow g\bar{g} + r\bar{r} \qquad\qquad g\bar{b} + b\bar{g} \rightarrow b\bar{b} + g\bar{g}$$

Nevertheless, for our purposes, each gluon can be thought of as having one color and one anti-color. Neither quarks nor gluons have been detected, but the experimental interaction products confirm the predicted octet shown above and the decays as shown in **section 10** and the *RPP.*

#13.4.2. QUARKS INTERACT WITH THE STRONG NUCLEAR (COLOR) FORCE GLUONS.

The strong nuclear force does not mediate electric charge. It instead mediates based on the color and anti-color of quarks and anti-quarks. It does not differentiate whether the quarks are up as in a proton or down as in a neutron. The different color combinations of the three colors of quarks and anti-quarks are allowed to be in one of eight different states arranged in eight different ways leading to the eight strong nuclear force gluons. Each gluon carries both color and anti-color charge. Each gluon is its own anti-gluon. Like the photon, the strong nuclear force gluons do not have mass.

#13.4.3. HADRONS ARE PRODUCED WHEN QUARKS INTERACT WITH GLUONS.

The hadrons are color-neutral composite particles. The hadrons are made up of various color quarks and gluons to make color-neutral mesons and baryons. The hadrons are held together by the strong nuclear force gluons. Quarks carry the strong force color charge in units of red, blue, and green just as particles such as electrons carry electromagnetic charge. Quarks also carry electric charges in units of plus or minus two thirds ($\pm2/3$) and plus or minus one-third ($\pm1/3$) that are mediated by photons and weak force bosons as explained in **sections 13.3** and **13.5.** Single quarks are not observable. Thus, the electric charge carried by quarks is only encountered in the form of combined quarks (mesons) with zero charge, or baryons such as protons with a charge of =1, neutrons with zero electromagnetic charge, or other baryons with a charge of -2, -1, +1, or +2.

Two quarks can interact with a mediator gluon to form a colorless meson. The color r and anti-color r cancel to form a colorless meson held together by a gluon(s). For example for a red r charm c quark and a not-red not-up quark,

$$c(r) + \bar{u}\,(\bar{r}) + g \quad \rightarrow \ (r)(\,\bar{r})\,c\bar{u}g$$

The gluon(s) is usually not shown in these type equations, but is always shown on Feynman diagrams.

A gluon can carry away blue and substitute r, then the red and anti-red quarks form a colorless meson held together by a colorless gluon(s):

$$c(b) + \bar{u}\,(\bar{r}) + g\,(r\,\bar{b}) \quad \rightarrow \quad c(r) + \bar{u}\,(\bar{r}) + g\,(b\,\bar{b}) \ \rightarrow \ c(r)\bar{u}(\bar{r}) + g$$

#13.4.4. STRONG NUCLEAR (COLOR) FORCE GLUONS ARE INFINITE IN VALUE AND LONG RANGE IN PRINCIPLE, BUT ARE USUALLY CONFINED WITHIN A HADRON (MESON OR BARYON). The strong nuclear force is infinite in value and long range like the electric charge but it is usually confined to short range within a hadron. Quarks inside the hadron are normally "loosely bound" and shield the gluon(s) holding the quarks together from having any influence outside the quarks.

If something happens to pull the quarks apart, the color force rapidly peaks to keep them together and then begins to slowly decrease with distance from the center of the hadron if the outside influence continues to pull the quarks apart. If the outside influence continues and finally overcomes the gluons, then the bond holding the quarks together breaks and the two (or three) quarks are momentarily free. BUT NOT FOR LONG! The energy of the encounter is now sufficient to create one or two pairs of virtual quarks (**section 12.7**) that gluons will immediately bond to the quarks that have been just torn apart. So if pair of quarks in a meson was pulled apart, there are now four quarks contained in two mesons. If there were three quarks in a baryon, there are now six quarks contained in two baryons. Unless a neutron or proton was formed, the mesons and baryons will almost immediately decay into more stable particles. The decay of all the particles continues until only stable particles remain (electrons, protons, neutrons, neutrinos, and photons).

Like the photon, the strong nuclear force gluons do not have mass. Like the photon which does carries electromagnetic charge, the gluons carry color charge. Gluons, however, can directly interact with and bond to other gluons to form color-neutral glu(e)balls) in two, three, and four gluon bindings. Like quarks, only indirect evidence of glueball existence has been observed.

Much like a photon, a single quark will have infinite color charge and send out an aura of mass-less color bosons of the appropriate color charge encircling the quark. The aura starts out to travel to infinity at the speed of light, but before it gets very far it will encounter one or two other quarks, virtual or otherwise, with which the rules of quantum mechanics allow it to combine. The virtual gluons are absorbed by the other quark, increase its energy, and emit a virtual gluon to push it toward the first quark. Note that gluons like photons have a J spin of +1 or -1, but unlike the photon, the gluon spin can have a third value 0.

Even though the other quarks may have the same repulsive electric charge (+2/3 and +1/3, or -2/3 and -1/3), the strong nuclear force virtual gluons overcome the repulsive electromagnetic force. The quarks now continue to exchange and absorb color-charge gluons to attract them and hold them together as a meson or baryon, and perhaps come together to form a neutron (+2/3, -1/3,-1/3) or proton (+2.3,+2/3, -1/3).

#13.4.5. WHERE DID ALL THE MASS OF THE PROTON COME FROM?

If the up u quark weighs 3 (MeV/c^2), a down d quark weighs 7, and a gluon weighs 0, how much does a proton consisting of udd quarks weigh? Is it 17, 13, or _? The correct answer is 938.272 MeV/c^2. How does this comes about? The extra (938 – 17 = 921 MeV/c^2) of mass comes from the enormous energy of gluon(s) that is converted to energy-mass as they bond the quarks together **(section 11.7.2)**. The amount of energy E of the gluon(s) involved is given by Einstein's equation $E = mc^2$.   If a way to liberate the enormous energy of gluons could be found, it would be revolutionary. So now that the proton or any object has "mass" what gives it the characteristics of mass including inertia and gravity? The answer is the Higgs fields, Higgs bosons, and perhaps the gravitons. How the Higgs fields do that is described in **section 13.7.**

#13.5. THREE WEAK NUCLEAR FORCE BOSONS W$^+$ W$^-$ Z^0 CONTROL PARTICLE INTERACTION AND DECAY.

#13.5.1. ALL FERMIONS (LEPTONS AND QUARKS) INTERACT WITH WEAK CHARGE.

All fermions (leptons and quarks) carry weak charge. Weak nuclear force bosons (W$^+$/W$^-$ and Z^0) mediate (transmit) weak nuclear forces between fermions. The weak nuclear force is associated with binding of particles together. The weak nuclear force is also associated with particle decay. The weak nuclear force bosons (W$^+$/W$^-$ and Z^0) are the only bosons that have mass; this causes the weak nuclear force to be short range and act over less that 10^{-17} centimeters. Two weak nuclear force bosons (W$^+$ and W$^-$) carry electric charge. In addition to electric charge, the weak nuclear force bosons also carry weak isospin charge.

Any interaction mediated by photons can be mediated by the Z boson. Like the photon, the Z boson has a J spin of 1.

#13.5.2. WEAK NUCLEAR FORCE BOSON W^+ W^- Z^0 INTERACTION AND DECAY EQUATIONS.

Muon neutrino and electron interaction (scattering) can be mediated by Z boson. (A muon neutrino hits an electron and changes its trajectory.) Note the conservation of charge.

$$v_\mu + e^- + Z^0 \quad \rightarrow \quad v_\mu + e^-$$

Any process mediated by the photon can also be mediated by the Z. For example, two electrons interact and repel each other: (Proton – proton interaction with each other mediated by Z^0 is similar with the two electrons replaced by two protons.)

$$e^- + e^- + Z^0 \quad \rightarrow \quad e^- + e^-$$

The reader might wonder which interaction would take place: the one mediated by the photon or the one mediated by the Z boson. Actually, there is a probability for each to occur and both will occur accordingly over many interactions. Specifically, which one will occur at any specific given time cannot be determined. To observe weak interactions only, neutrino interaction with uncharged particles must be studied which is very difficult.

As particles decay, as almost all particles do, their products will sooner or later (much less than a second) end up as stable particles or in a stable particle of a lower energy level – a neutron, proton, electron, photon, or neutrino. Even the neutron will be stable only if it ends up in the nucleus of an atom with a proton held together by gluons.

Decay of a muon: The muon and its charge are absorbed by a W^- weak boson and a muon neutrino; the W^- in turn decays into an electron and anti-electron neutrino. Note how charge is conserved at each step of the decay.

$$\mu^- \rightarrow W^- + v_\mu \quad \rightarrow \quad e^- + \bar{v}_e + v_\mu$$

A down d quark with a charge of -1/3 can interact with a W⁻(-1) boson in steps to form an up u quark with a charge of (+2/3). The W⁻ then decays into an electron and an anti-electron neutrino. The W⁻ absorbs (-1) in charge from d leaving it with a charge of (+2/3) and changing it into a u(+2/3) quark. The W⁻(-1) then decays into an electron e and an anti-electron neutrino.

$$d(-1/3) + W^- \rightarrow u(+2/3) + W^-(-1) \rightarrow u(+2/3) + e^-(-1) + \bar{v}_e$$

However, because a single quark cannot be found in nature, the above decay cannot occur except in concert with other quarks.

Put the down d quark together with another down d and an up u quark, to form an n udd neutron, then the above decay shows the decay of a neutron udd into an electron and a proton uud held together by g gluon(s):

The W⁻ absorbs the -1 charge changing the down quark into an up u(+2/3) quark. The W⁻(-1) boson then decays into an electron e⁻(-1) and a anti-electron neutrino $\bar{v}_e$. The three remaining quarks are held together by gluon(s).

$$n \rightarrow ddug \rightarrow du(+1/3)g + d(-1/3) + Z^0 \rightarrow du(+1/3) + u(+2/3) + W^-(-1)$$
$$\rightarrow du(+1/3) + u(+2/3) + g + e^-(-1) + \bar{v}_e \rightarrow duug + e^- + \bar{v}_e$$
$$\rightarrow p + e^- + \bar{v}_e$$

There are many, many other ways the weak force bosons interact particularly in particle decays. Refer to the **RPP** for a complete listing of experimental results through 2012. For a weak force, the weak force is very busy. It carries electric charge and can mediate the interaction of charged particle such as electrons and protons just as photons do. It can absorb and convey charge to charge one type quark into another. It participates in particle decay. It participates in atomic fission and fusion. One weak force boson can directly couple with other weak force bosons just as one gluon can couple with other gluons. Note as well that the W boson has a J spin of ±1.

#13.6. PARTICLE INTERACTION AND DECAY RULES.

As has been described before, most all the 38 elementary particles and over 283 composite particles interact and spontaneously decay into a very few stable entities, the electron, neutron, proton, photon, and neutrinos. The lifetimes (LT) and decay modes of the various particles are provided in **sections 9 and 10**. When particles interact and decay, they are subject to various rules and laws. Many of these have been provided previously. The following is a summary of the most important ones:

(1) Particles **continue to decay into the lightest (least mass possible) particle** unless prohibited by another rule. The proton is the lightest baryon. The electron is the lightest charged lepton. The electron neutrino is the lightest uncharged lepton neutrino. The anti-particle counterparts of these are likewise the lightest anti-particles that do not decay further.

(2) Energy and momentum are conserved in every interaction. (See **section 12.2.**)

(3) Charge is conserved in every interaction. W^+ and W^- bosons can bring charge to an interaction or absorb charge and carry it away.

(4) Color is conserved. Only quarks and gluons have color charge. Mesons and baryons are colorless. Strong force gluons can bring color charge to quarks or carry color charge away from quarks but the same colors that go into an interaction must go away from an interaction

(5) Baryon number is conserved. Each quark has a baryon number of 1, an anti-quark - 1. Each baryon has a baryon number of 3, and anti-baryon -3. If a baryon is made of four quarks or anti-quarks, the baryon number is adjusted to 4 and -4. Mesons and anti-mesons have a baryon number of 0.

(6) Lepton number is conserved. Each lepton has a lepton number of 1. The same number of leptons that go into and interaction must come out of the interaction. In an electromagnetic interaction, the same particle that went in comes out accompanied by a photon. In a weak interaction, if a lepton goes in, a lepton comes out but maybe a different one. The color force does not mediate leptons – leptons do not carry color charge.

(7) Quark flavor (u, d, t, b, s, c) is conserved for electromagnetic or strong interactions, but not necessarily for weak force interactions. (See **section 13.5.**)

#13.7. HIGGS AND GRAVITON BOSONS ADD MASS TO PARTICLES AND WEAK-NUCLEAR-FORCE BOSONS.

Detecting the Higgs boson in July 2012 at CERN, gave validation to the Standard Model (SM) of particle physics and changed the way mass is understood.

The Higgs Bosons H^+ H^- H^0 and gravitons provide a revised theory of mass. This revised theory of mass leads to the already familiar theories of mass, motion, and gravity as set forth by Isaac Newton's laws of motion (**section 2.3.5**) and Albert Einstein's Special and General Theories of Relativity (**sections 21 and 22**).

#13.7.1. PROPERTIES OF MASS.

Mass has the property of inertia (**section 2.3.5**). The fabric of spacetime has been described as molasses that gives inertia to mass by impeding an object moving through it. This is perhaps accurate for an object at rest that tries to change its acceleration and velocity. The molasses makes it very difficult to move through it and change velocity.

The fabric of spacetime on the other hand, only **impedes a CHANGE of velocity** and unlike molasses, **does not affect objects moving at constant velocity or freefalling**. (Freefalling objects may or may not be accelerating.)

More exactly, rather than like molasses, to understand how inertia acts in spacetime, think of an object such as your flat hand in a circular tub of water. If your hand is at rest, the water engulfs your flat hand and resists it moving around the circumference of the tub. As your flat hand moves against the water, the water begins to move around in the tub as well until finally the water is moving at the same speed as your hand. At that point, the water no longer resists the circular motion of your hand.

Now try to stop or increase velocity of your hand. The motion of the water continues to move around the tub at the same rate, resists the change in velocity of your hand, and continues to resist the change in velocity of your hand until the water again is moving at the same velocity as your hand or has stopped moving if your hand has stopped moving.

Now instead of your hand, us a Ping-Pong paddle. Notice how much "inertia" the larger paddle has than your hand.

The water accordingly demonstrates how the Higgs fields impart inertia and mass to particles and objects. The Higgs fields resist change in velocity of a particle or object adding inertia in proportion to its total mass.

Mass m is of three types: rest mass m_o, relativistic mass m_r, and energy mass M_e. The term mass usually encompasses all three types of mass, but the reader should be sensitive to the context in which the term "mass" is used. How much mass an object has can be measured as the inertia of an object. Inertia resists a change in velocity of an object - the greater the mass, the greater the resistance the Higgs fields exert on a particle or object to change in velocity. The Higgs fields change mass (inertia) of an object as its relativistic velocity and energy change.

An object cannot take on extra energy without having its mass increase at the same time – a hot kettle weighs more (has more mass) than a cold kettle. A compressed spring weighs more than an uncompressed spring. A large star heated by nuclear fusion converting hydrogen to helium has enormous potential, heat, and other energy which adds a substantial amount of energy-mass to it. This is explained by Einstein's Equation (**section 1.2.1**) for the equivalence of mass m and energy E:

$$E = mc^2 \qquad m = E/c^2 \qquad \text{Where c is the velocity of light.}$$

Mass is due to particles interacting with the four scalar Higgs fields. This interaction gives mass characteristics to a particle. This along with the recently detected Higgs bosons and (still theoretical) gravitons imparts mass and mass characteristics to all elementary particles (leptons, quarks, and weak force bosons) that have mass. In turn, the mass of the elementary particles becomes the mass of all the hadrons, atoms, molecules, compounds, planets, stars, and galaxies in our Universe that are composed of them. The Higgs fields and Higgs bosons also impart mass to the particles and virtual particles in empty spacetime (**section 12.7**).

#13.7.2. THERE ARE FOUR SCALAR HIGGS FIELDS IN THE FABRIC OF SPACETIME.

The fabric of spacetime is all pervasive containing among other things **four Higgs fields.** Spacetime acts in concert with the Higgs fields, Higgs bosons, and the as yet undetected gravitons. Gravitons, if they do exist, should be detected at CERN in the next several years. If not, it's back to the drawing board for many physicists.

NOTE: In the following sections, the discussion assumes the existence of the as yet undetected gravitons. Section 13.7.9 provides alternate explanations of gravity.

Finding the Higgs boson at CERN in July 2012 supports the existence of four theoretical scalar "Higgs" fields in the all pervasive fabric of spacetime of our Universe and perhaps the fabric of empty spacetime beyond our Universe.

A scalar field has a value (magnitude) at every point but not direction. In addition to these Higgs fields, the fabric of spacetime is the Infinite "empty" pervasiveness of space that is teaming with virtual particles popping in and out of existence and usually being quickly annihilated (**section 12.7**).

Two of the Higgs scalar fields are charged and two are uncharged. The two charged scalar fields give mass to the two weak force bosons W^+ and W^- that can carry charge.

One of the remaining neutral scalar Higgs fields gives mass to the weak force neutral boson particle Z^0.   The other remaining scalar field gives mass to the other elementary particles (leptons and quarks), atoms, and objects that have mass.

The mathematics involved describes these Higgs characteristics as four distinct mathematical scalar fields. A scalar field is described as a quantitative value of the characteristic of the field at every point in spacetime. The four Higgs fields should be thought of as four characteristics of the fabric of spacetime that is pervasive everywhere, rather than four separate entities. Whether there are four or more fields in spacetime with many different sets of characteristics that are described by many different sets of mathematics or one composite field that takes many different sets of mathematics to describe makes no difference.

For newly minted particles, the Higgs fields impart rest mass (energy) in concert with the energy of gluons and Higgs bosons. For objects at rest in Higgs fields, there is no effect. For objects moving or freefalling at constant velocity, Higgs fields have no effect. For objects otherwise changing **velocity and energy**, the Higgs fields and Higgs bosons increase (or decrease) their mass and inertia according to Einstein's Special and General Relativity (**sections 21 and 22**). For objects acquiring or losing any kind of energy (heat, kinetic, pressure, stress, etc) the Higgs fields and Higgs bosons increase or decrease mass of an object according to Einstein's General Relativity. For objects being accelerated or decelerated the Higgs bosons increase or decrease mass in accordance with Einstein's General Relativity.

#13.7.3. MATHEMATICAL DESCRIPTION OF FIELDS CAN BE DERIVED FROM LAGRANGIAN FORMULATION OF ENERGY AND MOTION.

In the 1960's, physicists were having difficulties in formulating the mathematics of the theory of the weak nuclear force. Experiments had shown that the weak force bosons (W^+, W^-, and Z^0) had considerable mass while the other bosons (photons and gluons) were massless. To introduce terms to account for bosons with mass to the theory ruined the symmetry of the theory and the equations did not work. The solution was to find a way to introduce the terms along with additional terms that restored the symmetry. These additional terms added new fields to the theory, the Higgs fields.

A mathematical description of quantum mechanics and fields can be formulated from the classical mechanics "Lagrangian" (L) formulation of energy and motion. The Lagrangian technique eliminates the need to consider forces of constraint in equations of motion and enables dealing only with scalars (such as energy) instead of multiple vector forces and accelerations. Physicists adapted this classical physics technique to particle interactions, not by any direct derivation but educated guesses.

The changes that finally resolved seemingly insolvable problems gave stunningly correct results at the quantum level. The adoptions of interest here are described in the mathematics of quantum electrodynamics, quantum chromodynamics, and quantum gravity. These quantum field theories are called gauge theories. A quantum theory Lagrangian describes the associated field(s) and the interactions of the associated particles with the field(s).

The quantum electrodynamics theory predicted the already discovered massless photon as the electric force particle (mediator). The quantum chromodynamics theory predicted three Dirac color fields (red, blue, and green) that interact with the eight vector fields of the eight massless strong color force bosons. Quantum chromodynamics was also a remarkable success.

The quantum theory of gravity did not fare so well. It predicted three massless weak force bosons that were named Goldstone bosons. Unfortunately, no weak force massless bosons had been detected so massless weak force bosons were highly improbable. When weak force bosons (W^+ W^- Z^0) were experimentally detected, they turned out to have considerable mass.

To the rescue, several brilliant and intuitive physicists (one named Peter Higgs) found a way to repair the Lagrangian for the weak force by adding a new field variable. By other mathematical constructs, the Lagrangian acquired an additional degree of freedom which then cancelled the Goldstone boson out. The theory then (hopefully) matched the real world of experiments. The theory predicted a new massive boson and showed how weak force bosons, quarks, and leptons acquired mass from the Higgs fields, Higgs bosons, and graviton bosons. Now that the Higgs boson has been found, quantum gravity theory can be further verified experimentally.

The theory of Higgs fields came about to eliminate the infinities, weightless weak force bosons, and other problems with the Standard Model of Particle Physics (SM). Previous approaches could not be evaluated without creating new problems. The concept of the Higgs fields solved these problems and provided a way to add mass and gravity to leptons, quarks, and the weak force bosons W+, W$^-$, and Z^0.

The theory of quantum gravity introduced a new way of understanding how mass comes about, describes gravity at the quantum and particle levels, and explains how particles interact with each other in a scalar gravitational field.

#13.7.4. HIGGS FIELDS AND BOSONS CORRECT THEORETICAL PROBLEMS.

There are specific Higgs fields to add mass to the W^+ W^- Z weak force bosons in a matter similar to that for the leptons and quarks. The other bosons (photons and gluons) are massless. So why do these weak force bosons need mass? At times they are a carrier of electric charge. This puts them in the same category as the other charged particles - the charged leptons and quarks that also carry electric charge. Mass seems essential to a charged particle so it can interact with other charged particles and exchange charge with weak force bosons.

The weak nuclear force bosons are massive. As a result, they operate at short range. In earlier theoretical formulations, adding the weak nuclear force bosons to existing theory for photons and strong force bosons did not work as it required the weak force bosons and Higgs bosons to be massless. By skill and insight, the theory was reformulated.

When finished (**section 13.7.3),** the gauge theory of gravity removed the mathematical anomalies. Mass was added to the weak nuclear force bosons to match the mass determined experimentally. In addition, the theory added mass to all the other elementary particles that were observed to have it by a pervasive Higgs field of the fabric of spacetime. And new bosons were predicted by the revised theory – the Higgs bosons and graviton boson. And, in 2012, the Higgs boson was detected ETAC at CERN, verifying the marvelous theoretical work done 47 years previously by Peter Higgs and his colleagues François Englert, Gerald Guralnik, Carl Hagen, and Tom Kibble.

The three Higgs bosons H^+ H^- H^0 have zero spin, zero electric charge, and no color charge. Like other bosons, they are their own anti-particles. They are very unstable, decaying in about 10^{-24} seconds. The Higgs bosons interact within the four Higgs scalar fields that are part of the fabric of all pervasive spacetime. The graviton is a theoretical boson that has not yet been detected. It is theorized to have a spin of 2 which is indicative of the complexity of its purposes. The Higgs bosons and gravitons carry the message of mass and gravity throughout the all pervasive and infinite fabric of spacetime.

#13.7.5. DESCRIPTION OF HIGGS FIELDS.

There are four Higgs fields. Three of the Higgs fields and Higgs bosons give mass to weak force bosons, leptons, and quarks but not to the massless electromagnetic force bosons (photons) and strong force bosons. One Higgs field gives rise to bundles of energy, the Higgs boson(s) that add or remove energy from particles as their energy-mass increases or decreases in response to various stimuli and changes in stimuli.

The Higgs fields consist of two charged scalar Higgs fields and two neutral scalar Higgs fields. **The scalar involved is the scalar value of the mass field E_m of an object's mass.** A Higgs mass field is a scalar having magnitude but not direction. (Vectors have magnitude and direction.) To a first approximation, the value of the Higgs scalar field at any point **r (a vector)** is the value is the object's total mass m (rest, relativistic velocity, and energy) divided by the square of the distance **r** from the object (m/r^2).

These Higgs fields are all pervasive characteristics of infinite spacetime. The Higgs bosons of the two charged Higgs fields of the fabric of spacetime give rest mass to the positive and negative charged W^+ and W^- bosons. The Higgs bosons of the neutral Higgs field of the fabric of spacetime give rest mass to the neutral Z^0 boson.   The Higgs bosons of the other neutral Higgs field work with the fabric of spacetime to give rest, inertial mass, relativistic-mass, and energy-mass to the elementary particles - leptons and quarks, and in turn to all the hadrons, atoms, and objects of our Universe. The Higgs bosons and graviton boson are the mediator quanta of these four Higgs fields.

Particles (leptons, quarks, and weak-force bosons) acquire mass from interactions with the Higgs bosons and the Higgs fields. When a particle is created in whatever way, the Higgs fields and the Higgs bosons interact to give the particle its due rest mass as specified in **sections 9 and 10.** If the energy-mass carried by the Higgs boson is excessive, then the excessive energy is radiated away by the particle.

The Higgs fields provide a "screening" which allows a particle's "effective" mass to be felt by other particles and measured by scientists. The screening of the mass of particles by the Higgs field fabric and fabric of space is very similar to the screening of the electrical charge of an electron or proton (**section 13.3.2**). In the case of mass charge, its value is also infinite and its aura of gravitons also reaches out to infinity. But the screening by the Higgs fields in the fabric of space reduces the effective value of the mass to that which is measured. The value of the effective mass (gravitational) field E_m or aura for a mass m at a distance r away is given to a first approximation by:

$$E_m = m/r^2.$$

If Higgs bosons (and possibly gravitons too) interact with a particle such as a proton to give it mass because the three constituent quarks and strong force gluons (bosons) don't carry enough, one or more Higgs bosons will provide the missing energy-mass. Any energy from a Higgs boson that is excessive will simply be radiated away by the particle.

This is analogous to a newly created electron carrying charge but not having charge until it interacts with photons. Until photons create an aura around the electron and interact with other charged particles, it doesn't exist as an electron – no photons, no electromagnetic field. The infinite energy of the electron's electromagnetic photon aura comes from its ability to draw energy from the infinite energy of infinite empty spacetime. This borrowed energy, however, is always paid back sooner or later as the energy of the electron's infinite aura eventually terminates in infinite empty spacetime – energy debt paid in full!

#13.7.6. HIGGS FIELDS TAKE ON SCALAR VALUES AT EVERY POINT AND CREATE GEODESICS IN CURVED SPACE.

An electron carries mass, but it doesn't have mass until it interacts with Higgs bosons and the Higgs field in the fabric of spacetime. Once it has gained mass, it can then interact with other particles that have mass: no Higgs field and no Higgs bosons – no mass. But mass is much different than electric charge. Mass requires a scalar Higgs field. No such field is required for electric charge. Photons easily take care of all the business necessary for electrically charged particles by the radiated electromagnetic aura of photons from the electron to interact based on Coulomb's law (**section 13.3**).

Not so with mass! A Higgs field has a complex job to do. Like the charge of the electron, mass charge of an object creates an infinite scalar ripple throughout the Higgs field **anytime the mass or location of an object changes.**

 This scalar ripple in the Higgs field extends to infinity and permeates all the fabric of spacetime. This scalar gravitational ripple of gravitons travels out at the speed of light. This gravitational ripple changes the scalar value of the infinite Higgs scalar field at every point around an object in accordance with the scalar total mass m of the particle and the distance r away from an object. (r^2 is a scalar. Distance away squared (r^2) is a scalar magnitude not a vector.)

This scalar mass value goes with an object everywhere it goes and changes the value of the scalar Higgs field at every point - mass here is the composite of rest-mass, relativistic/velocity-mass, and energy-mass.

The direction the particle moves is immaterial in the Higgs field as the scalar value moves in the direction of particle motion **reflecting the velocity and acceleration of the particle**.

Thus, the scalar gravitational value around a mass changes the effective value of the Higgs scalar field at every point around a mass all the way to infinity. The **Higgs field** does not transmit the force of gravity; it **reflects the scalar value of the sums of all scalar mass values of all particles and objects in the Higgs fabric of spacetime. It warps (curves) spacetime into GEODESICS by changing the scalar value of the infinite scalar Higgs field in spacetime in accordance with the mass of an object at that point in spacetime.**

The same process applies to every mass everywhere. Each mass has a scalar mass field value aura of gravitons around it that expands at the speed of light as mass energy and location change. The aura of each mass has a scalar value everywhere that affects other masses. **The Higgs scalar field then takes on a net value everywhere as the sum of all the scalar auras at that point. (**Scalars add: 1 +4 = 5 or 6 − 3 = 3.) The value of the scalar sums could be negative (negative gravity).

Around every point in the Higgs field, there is direction of maximum scalar value on a line (a four-dimensional geodesic tunnel) in curved spacetime. THIS IS THE GEODESIC THAT AN OBJECT WILL FOLLOW IN FREEFALL IN CURVED SPACETIME UNDER THE MOMENTUM IMPETUS OF EMITTED GRAVITONS. This is the "FORCE" of Gravity.

This geodesic is a "warp" (curve) in spacetime. Each object (mass) in this composite aura will move along this warp (curve)under the momentum of emitted gravitons in the direction of the maximum warp – the freefall orbit (geodesic in spacetime) that is predicted exactly by Einstein's General Relativity (**section 22**) or to a close approximation by Newton's laws of gravitational attraction. For massive bodies and/or objects of high energy content moving at velocities near the speed of light, Newton's equations are not adequate.

How does the Higgs field "compute" the geodesic path and how an object should travel along it under the momentum impetus of emitted gravitons? It's just the scalar value of the Higgs fields at any point in spacetime, just as an electric field is a property of an electron and proton.

#13.7.7. OBJECTS WITH MASS FOLLOW GEODESIC WARPS IN SPACETIME.

How is a poor lonesome meteor traveling through space induced to follow the Higgs field scalar geodesic path of the composite gravity curve of spacetime toward a massive object such as Earth. In the case of opposite electric charges, the electric force F_E of attraction was created by the interchange of photons pushing together the two objects of opposite electric charge q_1 q_2 a distance r apart with a strength determined by:

$$F_E = q_1 q_2/r^2$$

In the case of similar electric charges, the electric force F_E of repulsion was created by the interchange of photons pushing the two objects apart. It's different with Gravity.

"FORCE OF GRAVITY" F_G is an effect of mass and energy altering the geometry of spacetime (warping spacetime) into geodesic curves (paths) of the shortest distance between two points in spacetime. **Objects in a geodesic are in a freefalling orbit and follow the geodesic in which they are moving. Their motion in the geodesic orbit gives the appearance and effect of what is called Gravity. The emitted gravitons provide the momentum to accelerate or decelerate the masses accordingly.**

MASS AND ENERGY OF OBJECTS CURVE SPACETIME! SPACETIME TELLS OBJECTS HOW TO MOVE (SECTION 22). GRAVITONS ACCELERATE OR DECELERATE THE OBJECTS.

A geodesic has four dimensions including spacetime. A geodesic is like a curved four-dimensional tunnel in spacetime. The geodesic tubular warp guides an object's motion (if any), and provides inertia as objects change velocity. The geodesic does not impede a constant velocity or freefalling object in any way.   There is no up or down, or left or right to a geodesic tubular curve. It is just a tunnel or warp in spacetime. (Our sense of up and down comes from the gravity of the Earth that holds us on the earth.)

It's important to note, the **geodesic** connecting two masses in the scalar Higgs field **always "knows" certain information about one mass or a pair of interacting masses**. This information is critical in the Higgs fields controlling particle and object interaction:

- The scalar masses of the two particles (objects).
- Where the particles are located.
- The relative velocities of the two particles.
- The distance between the two particles.
- Changes in the acceleration and velocity of particles.

Thus, a geodesic tubular warp between two objects always senses relative distance, velocity, and acceleration between the two objects based on the relative values of the Higgs scalar fields between the two objects.

Two objects with mass, if starting at rest and unrestrained, will freefall (accelerate) as they approach each other at rates as determined by their distance apart, relative masses, and differential scalar Higgs fields. As the masses accelerate, the curved geodesic tubular warp alters their passage to impart the correct inertia, velocity, acceleration, and mass. Gravitons will interact to increase their energies and corresponding masses. As gravitons change their relative velocities and accelerations, their masses will vary. (Or, if the two masses decelerate when moving away from each other, their masses will decrease.) This is analogous to an electron in a laboratory gaining huge mass as it accelerates in a geodesic tubular warp toward a positively charged plate. The lab also gains mass as it accelerates toward the electron, but because the mass of the lab is so huge with respect to the electron, the change in the lab's motion and mass is not enough to be measurable.

The added mass comes from added kinetic energy and increased relative velocity. This energy and associated increase of mass is from the energy of Higgs bosons added to the objects. This is like the added mass (inertia) of a heated object that comes from added Higgs bosons.

#13.7.8.   GRAVITY IS MORE THAN AN ATTRACTIVE FORCE.

What is gravity? Gravity is more than just an attractive force. It is not like two objects of different electrical charge (+ and -) attracting each other. If you place a metal mesh or cage around one of the objects, the objects will be shielded from one another and not attracted. But there is absolutely no way to shield from gravitational "attraction" or more accurately gravitational "acceleration."

Stand on top of the Empire State Building. One-hundred or so floors shielding you from the ground will make no difference. Your acceleration to the Earth will be the same. Try any other experiment. THERE IS NO WAY TO SHIELD FROM GRAVITY. The characteristic of objects creating curved spacetime and geodesic orbits is a characteristic of spacetime called Gravitational attraction.

Try warping spacetime yourself - you will never look at a falling object or gravity the same: Hold an object such as a tennis ball out at arm's length. Better yet, climb to the top of a high building or mountain. Drop the ball....

Of course, it accelerates and falls to the floor or ground. Drop any other object, a lead ball, a wooden ball, a big ball, a little ball, a wad of paper, or a feather. If in a vacuum where air resistance in not involved, the warping of spacetime and impetus of gravitons will cause them all to fall exactly the same and hit the ground or sidewalk at exactly the same time.   This effect of the warpage or curvature of spacetime and impetus of gravitons in a geodesic is called **Gravity.**

Technically, the lead ball or other object is in orbit around the Earth when it is dropped. It just was dropped with poor initial conditions and crashed into the Earth.

The final details of the parts that Higgs bosons and gravitons play in gravity and mass will be tested and confirmed by science, perhaps by experiment at CERN in the near future! Warping of spacetime itself was recently tested experimentally with the results seemingly showing actual warping of the spacetime fabric.

215

#13.7.9. ALTERNATE THEORIES OF GRAVITY.

Any object that has mass and energy causes spacetime (four-dimensional spacetime) around it to warp (curve) into a geodesic. The closer to the object, the more pronounced the warpage. For two objects, the warpage between them joins to form a dumbbell-like appearance. Two objects, thus, have an aura of four-dimensional warped spacetime around each of them and a geodesic "tube" of warped spacetime between them.

Several alternative theories of Gravity are being studied.

(1) GRAVITONS. The most likely explanation of "The Force" of Gravity" is the theoretical (yet undetected) gravitons described in as part of the Higgs boson discussion in previous sections. Gravitons push two masses together by being emitted from the two masses so that the graviton momentum pushes the two masses toward each other in a geodesic tubular warp just like photons push unlike electric charges together. The graviton pops out of theoretical solutions of the Lagrangian and SU (8) Lie Algebras. Higgs formulations and string theories also predict the graviton. Gravitons may be detected at CERN in the next several years.

(2) GEODESIC TUBULAR SHRINKAGE. The geodesic tunnel in spacetime could take energy from empty space to warp (shrink) in length to move two objects together along a geodesic curve in spacetime. The geodesic tube shrinks together at both ends to accelerate the two objects toward each other at the velocities and accelerations mandated by the scalar Higgs field. At this point, the mathematics (of relativity and particle physics) describes exactly what happens.

This tube of warpage creates a geodesic path between masses, shrinks, and pulls them together at an accelerating rate. The rate of acceleration for each is determined by the other's mass and their distance apart. A large object such as the Earth accelerates another object toward it rapidly (faster and faster). A small object has some, but not much, acceleration effect on another object.

There are several more approaches by physicists to explain gravity **[Freeman]** :

(3) Shadow of Holographic Reality

(4) Rippling Mirage

(5) An Illusion

(6) Antimatter

(7) Twistor Space (where lines are points and points are lines)

(8) Heat

(9) Thermodynamic Mirage

(10) Entropy

#13.7.10. DETECTION OF HIGGS BOSON IN JULY 2012.

In July 2112, CERN announced detection what seems to be a Higgs boson. Its mass is 126 GeV/c^2 (126,000 MeV/c^2).

Finding the Higgs confirms the validity of the SM. Many theorists expect super-symmetric particles to be found later at about 8-9 TeV.

There are possible links between the Higgs fields and cosmic inflation of space the first fraction of a second of the universe after the Big Bang. Scalar Higgs fields might be responsible. The Higgs scalar fields may also be the energy of empty space.

To produce Higgs bosons, two beams of protons were accelerated in opposite directions to very high energies by the collider at CERN. The two beams were allowed to collide within a particle detector. Occasionally, although rarely, a Higgs boson is created fleetingly as part of the collision byproducts. It seems natural that a Higgs boson is a product of the high-energy collision between protons, as a high energy (mass) particle is necessary to account for the mass of a proton. See **section 11.7.2**.

Because the Higgs boson decays very quickly, particle detectors cannot detect it directly. Instead, the detectors register all the decay products and from the data the decay process is reconstructed. If the observed decay products match a possible decay process of a Higgs boson, it indicates that a Higgs boson may have been created. In practice, many processes may produce similar decay signatures. Feynman diagrams are prepared and probability calculations are carried for each of these. If the detector detects more decay signatures consistently matching a Higgs boson than would otherwise be expected if Higgs bosons did not exist, then this is strong evidence that the Higgs boson exists.

#14.2. DARK MATTER: 21 PERCENT OF OUR UNIVERSE; DARK ENERGY: 72 PERCENT; ORDINARY MATTER: 7 PERCENT.

Determining the composition of our expanding Universe is a problem. It is a big dark mysterious problem! The ordinary energy and matter that are detectable with telescopes and related instruments accounts for only about 7 percent of the amount needed to explain the motions of galaxies, stars, planets, and other objects of our Universe that can be "seen." The missing matter seems to be of two types: Dark (non-luminous and non-absorbing) Matter seemingly accounting for about 21 percent and Dark Energy seemingly accounting for about 72 percent of the missing matter. In March 2013, the Planck Satellite confirmed the values above, that dark energy was about 3 percent less than previously thought and ordinary matter is about 3 percent more.

There is currently extensive and intensive theoretical and experimental research regarding Dark Matter, Dark Energy, and Cosmology. According to the **2014 RPP, [Olsen, p. 368]**, "Experimental advances along these multiple axis could confirm today's relatively simple, but frustrating incomplete "standard model" of cosmology, or they could force yet another radical revision in our understanding of energy, or gravity, or the spacetime structure of our Universe. " The following discussion Dark Matter and Dark Energy extends the material of earlier sections of this book to include the origin and composition of Dark Matter and the origin and operation of Dark Energy even though there is not universal agreement in these matters.

#14.2.1. DARK MATTER (21 PERCENT OF OUR UNIVERSE).

Beginning in 1933, cosmologists determined that a spinning cluster of galaxies did not seem to contain enough mass to keep the galaxies within it from flying out. Later, they determined as well, that a spinning galaxy did not seem to contain enough mass to keep the stars within it from flying out. No matter how hard or how long they looked, they couldn't detect, see, or find the necessary missing mass – so called Dark Matter.

The effect of a spinning galaxy cluster and a spinning galaxy is similar to a spinning disc with children riding on and going around with it. If a child is not very near the center, the centrifugal force of the spinning disc will throw the child out from the center and off the disc. (Great fun!) With a spinning galaxy cluster, a spinning galaxy, and even the Earth spinning around the Sun, the force of attraction of gravity at the center must be enough to keep the galaxies, stars, and Earth from flying out and away.

#14.2.2. COMPOSITION OF DARK MATTER AND ITS ORIGIN.

Dark matter exists. It has been mapped. Dark matter causes star light to bend as it passes dark matter on the way to telescopes.

The search is on for the composition of the missing dark matter (mass). Maps of dark matter have been made by various techniques showing that it extends and permeates out beyond the center of a galaxy. The bending of light by the gravity of dark matter has been used to make maps of dark energy. Using measurements of the variations of the cosmic microwave background radiation (CMBR) and of the special distribution of galaxies, finds a density of cold dark matter.

Candidates as listed in the **RPP** for dark matter include primordial Black Holes (created by burned out stars), WIMPs (Weakly Interacting Massive Particles) that are theoretical but as yet undetected, and Axions that are theoretically postulated by string theory but yet undetected. The detection of WIMPS requires underground laboratories to protect against background contamination by cosmic rays. The experiments searching for the composition of dark matter are detailed in the **2014 RRP.**   (See **Bibliography**.)

There is another origin and composition of Dark Energy theorized below suggested by review of the time line of the Big Bang (**section 6.2**) and baryonic quark combinations (**sections 10 and 11**):

Shortly (10^{-7} seconds)   after the Big Bang, quarks and gluons formed from the super-heated (10^{14} K) Big Bang Plasma (**section 6.2**). This plasma created quarks and gluons, which in turn created quark composites. As far as is usually explained, the result was the creation of electrically-neutral neutrons. The neutrons decayed (after 880 seconds) and formed protons and electrons which then formed clouds of hydrogen atoms – the origin of stars. The protons will decay after 10^{32} seconds, but most all were immediately locked in atoms.

Look at the Big Bang timeline (**section 6.2**) and the quark combinations (**sections 10 and 11**) again. When quarks and gluons were formed 10^{-7} seconds   after the Big Bang, the high temperatures (10^{14} K) would have created not only a simple quark combination like the neutron, but also various other neutral quark combinations of potentially 6, 9, 12, 15, etc quarks and maybe bunches in between. Most of these quark combinations would have decayed in much less than a microsecond, but some massive combinations, perhaps including the massive top quark, likely had a long lifetime like the proton, survived, and formed what is now seen and measured as dark matter (**section 11.7.4**).

The huge gravity of the surviving quark combinations might have been strong enough to clump them together as dark matter of a galaxy, and may have been necessary for galaxy formation.

Will these Dark Matter quark combinations if they exist ever be identified or created in the laboratory? Possibly, it may be the temperatures and energies required are too extreme. However, physicists are a brilliant, wily, resourceful bunch when it comes to challenges – it took over 40 years to finally create the Higgs boson ETAC in the collider at CERN....

#14.3. DARK ENERGY (72 PERCENT OF OUR UNIVERSE) IS CAUSING OUR UNIVERSE TO EXPAND AT EVER-INCREASING RATE.

One way to find the total mass and energy in the universe is to determine the rate of deceleration of the expansion of the universe. Edwin Hubble had determined that the Universe was expanding in 1928. At that time, scientists expected that the continuous processes in Space would overcome negative gravity (**section 6.2**), our Universe would gradually decelerate (stop expanding), and the force of gravity would ultimately cause all the objects and energy in the Universe to collapse together again as they were just before the Big Bang. Scientists set up experiments to find out for sure.

These experiments found that the Universe is expanding, but galaxies themselves are not expanding. Attractive gravity seems to hold each galaxy together. The distance between galaxies is expanding. It's like small coins pasted on a balloon. As the balloon inflates, the coins get farther apart, but the coins do not get any bigger.

The astonishing conclusion of various experiments in 1998 is that unexpectedly the expansion of our Universe expansion is not decelerating. Our Universe is not only continuing to expand, but the rate of expansion is accelerating. Universe expansion seems to have decelerated due to gravity of the Universe for the first 7 billion years or so after the Big Bang 13.8 billion years ago and then has unexpectedly accelerated at an ever-increasing rate.

The galaxies are not being pulled apart. The galaxies are being pushed away from each other in all directions. The Universe is being pushed apart. Galaxies in the Universe are moving away from each other. This type of expansion is what is to be expected as being caused by the Big Bang (**section 6.2**) where the velocity of expansion increases with distance from the origin of the Big Bang. The rate of expansion of our Universe slowed at first, slowed by attractive gravity when the galaxies were closer together, but is now accelerating as the galaxies are moving further apart.

A clue to what's going on is the negative energy of emptiness (empty space?) that was outside the Big Bang sphere. Negative energy is created by the negative pressure of empty space (**section 2.10**). The Big Bang sphere held extreme positive energy (high temperature and high positive pressure) inside it (**section 17**).

Ultimately, the Big Bang Sphere could no longer hold the built-up and pent-up positive pressure and positive energy (**section 17)** and it exploded with a **Big Bang.**

The unfurling (expansion) of our Universe is much like highly compressed gas in a tank **inside** a balloon suddenly being released into an enormous, a highly evacuated space. The highly compressed gas causes the balloon to be blown up quickly due to the negative pressure of the highly evacuated space (**section 2.10.3**). (If the pressure of space was the same as that of the compressed gas, the balloon would not be blown up.)

At the earliest moments of the Big Bang, the enormous energy of negative pressure of empty space outside the expanding Universe **rapidly** inflated (blew up) the Big Bang sphere in all directions around it. (This resulted in the conversion of the negative energy of space to positive energy of extremely high-temperature plasma (**section 6.2**) to fill the expanding Universe. This subsequently formed all the gas clouds, galaxies, stars, planets, and other objects in our Universe.

Einstein's original equations implied that the Universe would expand. In disbelief, Einstein made changes. Then, when Hubble found that the Universe is expanding, Einstein realized that he had missed an opportunity of a lifetime to predict the expansion and reversed his changes. Now scientists realize that the Universe is not only expanding, it is expanding at an increasing rate. They realize as well that further changes to Einstein's equations are necessary to incorporate the effects of negative pressure beyond our Universe or whatever dark energy is causing the accelerating expansion of our Universe.

The hypothesis herein is that the negative pressure of very cold (10^{-30} K) empty space beyond our universe is the Dark (negative) Energy that is causing our Universe to expand at an ever increasing rate. Many physicists believe, however, that there is nothing beyond our Universe.

The first spacecraft to ever leave our solar system, left in March 2013. It should not find anything untoward in its travels within our galaxy, the Milky Way. The real interest is when a space vehicle leaves our galaxy and travels in expanding space between galaxies. It may or may not find Dark Energy. Whatever it finds will put a lot of theoretical physicists to work. Things will get even more interesting when our ancestors are able to send a spacecraft beyond our Universe where it may be able to travel at warp-speed, much, much faster than the velocity of light (**section 12.7.2**).

#15.   DO BOTH ANTI-PARTICLE AS WELL AS PARTICLE UNIVERSES (SUCH AS OURS EXIST)?

If particles and anti-particles annihilate each other, why is our Universe made up of mostly particles and only rarely are anti-particles encountered? A possible explanation is because when our Universe was formed preference was somehow given to particles not anti-particles.

 Certain mesons and anti-mesons have been found to decay slightly differently into kaons (**section 10.2**) which then decay further under the weak nuclear force bosons giving a very slight preference to particles over anti-particles. The mathematics of this decay is explicit. The slight difference, however, is not enough to explain the almost complete absence of anti-particles and anti-matter in our Universe. The theory of super-symmetric particles further supports this premise and detecting super-symmetric particles at CERN, will provide additional and perhaps conclusive evidence that our Universe was naturally formed as a positive Universe and all others will also be formed as positive universes.

Reasoning based on our Universe alone leads to the hypothesis that negative universes (with a preponderance of anti-particles) cannot exist, only positive universes such as ours (with a dearth of anti-particles). Extending this hypothesis is (a giant leap). Nevertheless, other universes probably will be much like ours from a geological and cosmological standpoint, and will obey the same laws of physics that ours does.

So at least for now, it seems likely that besides our positive Universe, only positive universes just like ours existed, exist, or will exist elsewhere. Not every physicist agrees with this hypothesis.

It may turn out that which type of universe will be created is a "symmetry" issue. Several symmetry issues exist in particle physics. In a symmetry issue, sometimes two or several outcomes are equally likely and the choice made by our Universe is simply a matter of chance.

To answer these questions, it will be necessary to send a space probe beyond our Universe. Fortunately the first probes will be unmanned, as no one can know for sure what lies beyond.   It seems likely that empty space beyond our Universe is just like empty space unfolding in our Universe, and universes there will be just like our positive Universe made up of galaxies, stars, suns, and planets.

 String theory is a potential theory of everything. It is mathematically self-consistent from the quantum world to the cosmological world. In its current state of development, some string theory theorists believe that there may be as many as 10^{500} different kinds of universes and that all but one (ours) may obey different physical laws than ours. To calculate and test each of these will take many, many lifetimes and is not currently feasible or possible unless a breakthrough of some sort occurs such as happened when J. Craig Venter discovered a rapid way to decode the human genome and reduced the time required from decades to months.

#16 WHAT ABOUT OTHER KINDS OF UNIVERSES?

There are probably other kinds of universes out there besides our Universe.
Theorists postulate various kinds of universes such as bubble, membrane, parallel, and so forth.   Theorists make statements about these theoretical universes, such as exact copy of this Universe may exist with exact copies of all the people alive right now in it. Perhaps, there is to a probability of 1 chance in $10^{1,000,000,000,000,000,000,000,000,000,000}$ or so. (Okay, maybe that's an exaggeration.) So far, not a sign of any of these (particularly with another me in it) has shown up. But based on the last 100 years of developments in particle physics, cosmology, and relativity, the best is still ahead!

Nevertheless, there are probably other universes out there besides our Universe. See **section 17** below.

#17. PUTTING IT ALL TOGETHER – WHERE AND HOW DID BIG BANG COME ABOUT BEFORE OUR UNIVERSE EXISTED?

Somehow all the elementary particles, spacetime, and the Higgs fields in the fabric of spacetime came about when our Universe was created by a Big Bang. **Section 6** provides detailed information about what happened **AFTER** the Big Bang went "**BANG!** There is general agreement among physicists and cosmologists of what happened and and how it transpired.

Perhaps the first **section** of this book should have explained what went on before the Big Bang, what caused it to go "**BANG!,** and where it took place as well. There were practical reasons to delay it in this book until now.

Theoretical models (mathematical and otherwise) of the Big Bang sphere are difficult to formulate. The Big Bang sphere was created before it "exploded" into the Big Bang. No direct experiments are possible. All the conclusions about the creation of the Big Bang sphere are based on indirect evidence and theoretical models. The temperatures and pressures associated with creation of the Big Bang approach infinity. Infinities cause inconsistencies, anomalies, and other problems in theories and mathematical formulations particularly those based on the elementary particles (leptons, quarks, and gluons).   To resolve some of the problems, the creation of the Big Bang is described in terms of vibrating energy strings – this will be done in the next several **sections.**

Some physicists and cosmologists believe that there was no Time or Space before the Big Bang – Time and everything started at the Big Bang. It has been said that asking what happened before the Big Bang is like asking what is north of the North Pole. (Of course, there may be nothing north of the North Pole, but there is a lot of space junk above the North Pole.)

Sections 2.10.3, 6.2. and 14.2.2 provide a description of the of Big Bang from various viewpoints. Very near the beginning of the Big Bang (10^{-35} seconds after the Big Bang) a huge 10^{30} inflation (expansion) of our Universe took place that lasted less than a trillionth (10^{-12}) of a second. This inflation was caused by the high temperature and enormous pressure of the Big Bang sphere suddenly expanding (exploding) into the infinite negative pressure and infinite negative gravity of the "emptiness" of empty space beyond our Universe that our Universe was unfurling into.

The energy of this inflation into the infinite negative pressure and gravity beyond our Universe in turn created all the energy and mass of our Universe **(section 6.2).**

Here's the problem created by this inflation for physicists. The rapid inflationary expansion of the new Universe caused it to expand faster than the velocity of light! Now everyone knows that nothing can travel faster than the velocity of light! However, physicists came up with a "workaround" explanation: If the new Universe was expanding at a velocity greater than the velocity of light, but was not expanding into something, and instead was just unfurling into "nothingness" at greater than the velocity of light, that is okay! This expansion could happen at greater velocity than the velocity of light yet would not violate the principle that nothing could happen at a velocity greater than that of light.

Based on this, many physicists believe that there is no time, no nothing outside our Universe, except maybe more universes **(section 16).** Our Universe is expanding **(section 6)** into nothingness!!!

However, it is the intention of this book not to let the matter rest there, but to provide another viewpoint for the reader in the rest of **section 17.**

#17.1. QUANTUM STRINGS OF VIBRATING ENERGY.

Energy quanta are the vibrating strings of pure energy alluded to at the beginning of this book **in section 2.2**. These vibrating strings of pure energy make up the fundamental particles (leptons, quarks, and bosons) described in **section 9** from which all atoms, molecules, and objects in the Universe are made. It is assumed here that every fundamental string is exactly like every other. Also, these fundamental strings create all of the fundamental particles (**sections 9 and 10).** To do this, **a fundamental string by itself or in conjunction with one of more identical fundamental strings vibrates (oscillates) as a stationary wave open or closed (in a loop) in different modes of oscillation**. **Section 3** provides an overview of oscillating sinusoidal waves, wave motion, wave cancellation, wave reinforcement, and wave interference.

The quantum energy strings are theorized to vibrate in 11 different dimensions. Four dimensions are the ones of our everyday perception - up/down, right/left, forward/backward, and time (or x, y, z, and time T). The other seven dimensions are best described as mathematical contrivances (such as spin and color in the theories of the Standard Model), that may or may not have physical characteristics or dimensions that are real, but are necessary to make the mathematics work and give accurate results that match experiments. Mathematicians and physicists often work in mathematical spaces of infinite dimensions and use mathematical contrivances that turn out to match experimental results with astounding accuracy.

231

#17.2. FABRIC OF EMPTY SPACE THAT EXISTED BEFORE BIG BANG CREATED VIRTUAL PARTICLES AND VIRTUAL ANTI-PARTICLES – THESE LED TO SPHERE OF BIG BANG.

Section 12.7 explained how empty space is not empty. Empty Space is teaming with virtual particle pairs and energy. These virtual particle pairs in our Universe are leptons, quarks, bosons, and their anti-particles (**section 9**). They could also be virtual vibrating strings and anti-strings." They are created by borrowing energy from empty Space, usually exist for a short time, and then disappear being annihilated with their anti-particle partner in a puff of energy - returning the borrowed energy to empty space. But even though they are mostly short lived, they give the supposed "empty" fabric of Spacetime, a background of virtual particles and energy that can be measured (**section 12.7**).

The fabric of empty spacetime in and beyond our Universe can be thought of as a four-dimensional fabric that stretches in all directions to infinity - four dimensional, as the fourth dimension is time. (See **section 21**.) It's out of this fabric of spacetime that the virtual particle pairs spring into existence by borrowing energy from the fabric of spacetime.

The borrowed energy must be paid back in full and the fabric of spacetime will get its due sooner or later, rest assured. So for every "cup full" of positive energy borrowed from spacetime fabric, there is a cupful of negative energy left behind to be refilled later.

But, as the locations of any quantum particle and anti-particle have a probability (**section 12.4.2**) of being anywhere, sometimes one of the virtual particle pairs does not get reunited with its mate (anti-particle partner). This unfaithful particle or anti-particle is fully equipped with all its attributes whether an anti-particle pair of the neutrino, electron, quark, string, or other particle. A virtual quark may be unfaithful to its anti-quark partner and take up with another forlorn anti-quark particle and form a meson. Three forlorn quarks may form a baryon. A forlorn proton baryon may take up with an abandoned electron and form a hydrogen atom.

All of these unfaithful particles leave empty energy holes behind that sooner or later will be filled. So how did these virtual particles in infinite empty space conspire to create a Big Bang? One possibility is described below. A more likely way is described in **section 17.4.**

As these virtual particles have energy, they also have mass and gravity. Over eons, the number of particles being attracted together by gravity into a sphere by the warping of spacetime (**section 22**) becomes enormous. Gravity takes over as the strongest force as the amount of collected mass becomes a sphere, like a gigantic wave that builds and builds out of other waves due to freak conditions until it is gigantic and comes crashing over a ship or a small island reeking great loss of life. With gravity, the dense sphere of virtual mass material gets denser and hotter, and the outward (positive) pressure gets higher as the amount of mass accumulates.

The warping of space at the dense sphere gets so extreme that all nearby virtual particles (or anti-particles) being created are attracted. Ultimately, the amount of mass in the sphere makes it so dense and hot that it cannot contain itself and a Big Bang occurs.

However, there is a problem with this approach of creating a Big Bang by combining particles into a sphere by gravity! As the sphere gets denser and hotter, instead of making a Big Bang it may end some other way (section 17.3).

#17.3. SPHERE OF VIRTUAL LEPTONS, QUARKS, GLUONS, AND BARYONS MAY NOT BE ABLE TO OBTAIN HIGH ENOUGH PRESSURE, TEMPERATURE, AND DENSITY TO CAUSE BIG BANG.

It may not be possible for a sphere of virtual leptons, quarks, gluons, and baryons to get out of control and arrive at high enough pressures and temperatures to cause a Big Bang:

(1) A sphere made of hydrogen or other elements would just act like a star as pressure increases by igniting and burning by nuclear fusion or fission. In some cases, a black hole will be created.

(2) A sphere made of material like a planet will have a hot core that causes earthquakes and eruptions. In some cases, a black hole will be created.

Mass of electron neutrino 1×10^{-6} MeV = 1 eV = 1.6×10^{-19} joules
Diameter/Mass of Electron 1×10^{-16} cm   0.5 MeV = 0.8×10^{-13} joules
Diameter / Mass of Proton 0.8768×10^{-13} cm   938 MeV
Diameter of Hydrogen Atom 1×10^{-8} cm
Diameter o f Quantum String 0 cm (?)
Length /Mass of Quantum String 1×10^{-33} cm 1×10^{-9} eV = 1.6×10^{-28} joules
Diameter of Planck Sphere 1×10^{-33} cm
Diameter of Big Bang Sphere 1×10^{-33} cm (?)

(3) At near 0 K (10^{-30} K) temperatures, it may be possibly too cold for virtual leptons, quarks, bosons, and baryons and their anti-particles to spring into existence. They may require too much energy.

Another way of forming the Big Bang Sphere is given in section 17.5

#17.4. COMBINING OF VIRTUAL PARTICLES TO CREATE BIG BANG IS DETERMINED BY PROBABILITIES.

Is the combing of particles into a Big Bang sphere by quantum probabilities even possible?   Yes!! **See sections 12.4 through 12.4.3**.

That's like asking if throwing boxcars with a pair of dice is routine. Yes, it is routine to throw boxcars about 1 time in 36 rolls of the dice. No, it's not routine to throw boxcars 10 times in a row, or 1000 times in a row, or a million times in a row, a trillion times in a row, or 10^{50} times in a row. No, each of these is not routine. Is each of these possible? Well yes, but highly improbable. But infinite time is a long, long time – anything that can happen, **will happen**, but not very often.

Infinite empty space has a probability of creating enough virtual particles to create a Big Bang, putting them all in exactly the same place at the same time, and creating a Big Bang sphere ready to go BIG BANG almost instantly.

It is much more probable, however, that a Big Bang will be created by quantum vibrating strings rather than virtual leptons, quarks, gluons, and baryons. See **section 17.5**.

#17.5. BIG BANG WAS CREATED BY QUANTUM VIBRATING STRINGS.

Start with empty space somewhere long (13.8 billion years) ago and far away devoid of any traces of a Universe - ours or some other one. Assume the temperature there is very, very near absolute zero Kelvin degrees (10^{-30} K). (Temperature of empty space in our Universe is about 2.15 K. The lowest temperature on Earth 10^{-9} K was reached in a laboratory recently.)

At near 0 K temperatures, it's too cold for virtual leptons, quarks, bosons, and baryons and their anti-particles to spring into existence. They require too much energy. However, virtual strings require much less energy and along with their virtual anti-strings (which are exactly 180 degrees out of phase) can pop into existence. (See **section 17.1.**)

Now assume an empty Planck sphere of diameter 10^{-33} cm. **Section 17.3** states that the length of a quantum vibrating energy string is also 10^{-33} cm. Assume that the wavelength of the quantum vibrating energy string waves is such that it is confined to a **Planck sphere of diameter 10^{-33} cm.**

Now almost instantly, fill this sphere of quantum vibrating strings with enough energy to make a Big Bang as described **in section 17.4.**   How many strings can be put in the Planck sphere? Answer: **INFINITE! (Well, anyway more than enough to make a Big Bang.   See section 17.7.)**

#17.6. AS TEMPERATURE AND PRESSURE INCREASE IN BIG BANG SPHERE, BIG BANG OCCURS.

Temperature and pressure increase almost instantly in Big Bang sphere (**section 17.5**) from increased numbers of resonating vibrating strings.

A Big Bang occurs with enough pent up mass, heat, and pressure (energy) to give birth to an entire universe with a self-sustaining, inflationary, mass-creating expansion.

From this point on, things progress as stated in **section 6 and 14.2.**

#17.7. HOW MUCH NET ENERGY DOES IT TAKE IN BIG BANG TO MAKE OUR UNIVERSE?

Our Universe has a lot of **POSITIVE ENERGY**:
- 4×10^{11} (400 billion) galaxies
- 4×10^{23} stars (1 trillion stars per galaxy)
- Planets, comets, asteroids, and other contents
- Dark energy (72 percent of our Universe)

To calculate the energy E of the Universe:

Mass of Universe = **1×10^{50} kg**

Additional energy-mass of Universe = 1×10^{20} kg
(Due to heat of stars, gravitational energy, potential energy of unburned helium, etc.)

Total equivalent mass of Universe = 1×10^{70} kg

Using Einstein's equation, E= mc^2

$$E = (1 \times 10^{70} \text{ kg}) \times (3 \times 10^8 \text{ m /sec })^2 = 9 \times 10^{86} \text{ kg m}^2/\text{sec}^2$$

$$= 9 \times 10^{86} \text{ joules}$$

Total Positive Energy of Universe = 9×10^{86} joules

There are several other ways to calculate the total energy of the Universe based on:

- Number of stars in the Universe (1×10^{23}).
- Number of galaxies in the Universe (3×10^{11}).
- Number of protons in the Universe.
- Energy density of the Universe.
- Volume of the Universe.

The results using these methods and the one used above range from about 10^{50} to 10^{90} joules. A mid range of 10^{70} joules is perhaps reasonable.

One other way of computing the total mass of the Universe is to start with the

Energy Density of Universe = 10^{-23} gm/m^3 (about 5 hydrogen atoms per m^3)

The volume of the Universe is 1.075×10^{72} m^3

The total mass of the Universe is $(1.075 \times 10^{72}$ m^3) x $(10^{-23}$ gm/m^3) = **1.075×10^{49} gm**

So the calculations by two of the several ways for mass of the Universe lead to relatively close results.

Now consider the **NEGATIVE ENERGY** in our Universe.

Section 2.10.3 describes the almost infinite **negative** energy of empty space before the Big Bang.

All the positive energy of the Universe came from the negative energy of the empty space. It takes negative energy to create positive energy. Positive energy cannot be created otherwise; otherwise empty space would be unstable!

Right now, about 9×10^{86} joules of negative energy has been borrowed from the negative energy of empty space to make its positive energy!

As a result, when all the **positive and negative energy** in our Universe are considered, and all the evaluations are complete, the result is the positive and negative energies in our Universe balance out.

The net amount of positive and negative energy is zero ●

Therefore, the net energy required to create our Universe is zero ●

#17.8. VIBRATING STRINGS INITIATED BIG BANG WITH LESS PRESSURE, TEMPERATURE, AND ENERGY THAN EXPECTED.

A Big Bang can be initiated with a lot less energy, heat, and pressure, and a lot fewer vibrating strings than is expected based on the analysis of **section 17.7.** First, as explained **in section 6.2**, when the Big Bang is initiated, there is an enormous 10^{30} inflation in 10^{-35} second. This inflation is caused by the negative pressure and negative gravity **section 2.10.3** of empty space (**section 12.7).** It introduces enormous energy (energy of inflation) into the newly expanding Universe creating a huge positive energy – energy that does not have to be provided by vibrating strings in the Big Bang sphere.

Second, in the case of the big Bang sphere of vibrating strings, one other factor that reduces the number of vibrating strings required to a viable amount is the principle of resonance. Resonance of vibrating strings occurs when the vibrations of one string excites the other strings to vibrate. Then each vibrating string incites all the others to vibrate, building to a huge crescendo of enormous energy causing the Big Bang.

 This is similar to a bridge designed to handle tremendous loads being subjected to moderate traffic or winds that cause it to vibrate at its **resonate frequency**, tearing the bridge apart. Bridge designers accordingly design a bridge that cannot be placed in a resonate mode. Sound directed at a cavity can cause the cavity to resonate (reverberate) and greatly increase sound amplitude. Soldiers marching in step across a bridge can resonate with the bridge and cause it to collapse. A singer is able to break a glass with his (or her) voice by holding a note at exactly the resonate frequency of the glass.

Think of the quantum vibrating strings vibrating at exactly the resonate frequency of the Big Bang sphere and causing enough "peak" pressure to cause it to explode with a Big Bang and create our Universe.

Once the Big Bang has been initiated, **section 6.2** describes creation of the Universe.

#17.9. ENERGY BORROWED FROM EMPTY SPACE TO CREATE UNIVERSE MUST BE REPAID.

What of all the holes of negative energy left behind after the Big Bang? They must be paid! The energy must be returned.

Empty space must be returned to its original condition! Since energy is mass, the empty energy holes may have migrated together by gravitational attraction into one or several very huge holes of negative energy (negative gravity) out in spacetime beyond our Universe, that demand their due. Likely, this huge reservoir of negative energy is the dark energy causing the expansion of our Universe to accelerate just like gravity causes increasing acceleration. Ultimately, our Universe will continue to expand, age, and grow cold as all stars burn out, and all remaining matter and energy is absorbed back into the energy holes in spacetime fabric. Borrowed energy paid back in full!!!

How often does a Big Bang occur? Not very often, as no trace of another universe has been found. Perhaps there has never been another universe before ours. Perhaps there will never be another after ours. More likely, universes are created regularly probably taking trillions of years to form. That doesn't preclude more than one universe existing at the same time. Perhaps someday, there will be intra-universe travel.   Some physicists have developed mathematics that show that our Universe was formed on a "bubble" of the Big Bang and that many other universes were created at the same time on other bubbles. Other physicists have derived mathematics for parallel and mirror universes. There may be mathematical conjecture; but there is no experimental evidence for multi-universes. Assuredly though, there is experimental evidence for the existence of our Universe.

#17.10. PUTTING EVERYTHING TOGETHER TO FROM BEGINNING TO END OF OUR UNIVERSE. The previous sections in this book as referenced below provide the basic information necessary to understand how our Universe was created at its beginning. The beginning of this section provides a synopsis this basic information. This synopsis is followed by expanded information about the big bang in **sections 17.11 through 17.11.8**. **Sections 17.12 through 17.13** describe how our Universe will end.

REFERENCES: Sections 9 and 10. The elementary particles in our Universe were created by the Big Bang and interact with each other to form composite particles.
REFERENCE: Section 6. The composite particles form all 92 natural elements (hydrogen, helium, …, gold, …, uranium) and 23 manufactured elements (Neptunium, …, Lawrencium), 115 elements in all. The 92 natural elements in our Universe make up all its galaxies, stars, planets, etc.

REFERENCES: Sections 12.2.1 and 13.6. Key particle conservation characteristics govern particle interactions: energy, momentum, charge color, baryon number, lepton number, and quark flavor.
REFERENCE: Section 12.3. All particles and objects exhibit both particle and de Broglie wave characteristics. For large objects (cars, moons, planets, elements, composite particles, etc), particle characteristics predominate. For small (quantum) objects and elementary particles (quarks, electrons, photons, etc), wave characteristics predominate in many instances. For example, an electron acts as a particle in many instances, and as a wave in an element or by itself.

REFERENCE: Section 12.4.5. The Uncertainty Principle makes it impossible to determine both a particle's location and energy (velocity) at the same time. Both the wave and particle properties of matter cannot be observed at the same time.
REFERENCE: Section 12.4.2. Particles have probabilities of being at any specific location at any specific time. If a particle has a probability of being at several different locations at any specific time, the sum of all the probabilities must add up to one: 1.00. The probability characteristics of a particle are the particle. When a particle has selected a specific location, the probabilities of all the other possible locations immediately reduce to zero.

REFERENCE: Section 12.4.3. If a quantum particle or wave has several or many possible paths which could be traveled, the particle or wave exists and travels in all possible paths at once. The particle or wave ends up taking one of these paths only when something happens to force the particle to select one of as many as billions of possibilities. Over many, many particles or waves, the actual results can be predicted to a high degree of accuracy by the calculations of quantum mechanics.

REFERENCE: Section 12.6. If two or more quantum particles are entangled, if the state of one of the entangled particles is arbitrarily selected, all the other associated entangled particles **IMMEDIATELY** are forced to select their required state, no matter how far the entangled particles are separated, a nanometer or light years: **INSTANTLY**, faster than the speed of light. Research in this area may lead to instantaneous communications anywhere in the Universe.

REFERENCES: Sections 12.7 through section 12.7.2. Empty Space in our Universe and Empty Space beyond Our Universe Is Not Empty. Empty Space is teaming with various quantum virtual particle pairs and virtual wave pairs. These particle and wave pairs come into being by borrowing paired characteristics (such as + and - electric charge) from Empty Space. This borrowed energy and electric charge of particle characteristic pairs must be ultimately repaid. If the two members of these pairs later interact, they will exactly cancel (annihilate) each other and return the borrowed energy and paired characteristics to Empty Space. One member of a pair of virtual particles or virtual waves can interact with the member of a similar pair leaving behind its partner pair to fend for itself until another member of a similar pair comes along to annihilate it.

REFERENCE: Section 14.3. Our Universe came about by borrowing energy and particles from Empty Space. Dark energy is causing our or Universe to rush apart to pay back the borrowed energy and particles. **Empty Space beyond our Universe is Energy and Charge Neutral. It is also a source of an infinite amount of Energy and particle characteristic pairs which can be borrowed but must be repaid.** Our Universe is now rushing to repay the Energy and paired characteristics borrowed by the big Bang as Dark Energy causes our Universe to expand in all directions.

#17.11.   QUANTUM WAVES AND STRINGS OF VIBRATING ENERGY ARE ESSENTIAL FOR A "BIG BANG" TO CREATE A UNIVERSE.

To simplify Quantum String Theory (which is under development and postulates numerous types of vibrating strings), the following assumptions are made to give an intuitive explanation of how vibrating strings created a Universe by a Big Bang and enable computing the associated probabilities:

#17.11.1. Assume that there are four types of vibrating strings: Red (+), Green (-), White (1), and Black (-1). Two of these create and carry plus (+) and minus (-) electric charge as indicated. (It may turn out that there are many more types of vibrating strings that must be taken into consideration in this type of analysis and conjecture - there are likely nine different types of strings, one for each elementary particle.)

#17.11.2. These four types of strings RGWB are virtual strings that spring into existence RANDOMLY and exist until ANNIHILATED by their partners: R by G, and W by B.

#17.11.3. Assume that the following sequence of these strings is the only none-self annihilating sequence:

RWGB RWGB RWGB RWGB ….
WGBR WGBR WGBR WGBR ….
GBRW GBRW GBRW GBRW ….
BRWG BRWG BRWG BRWG ….

….. ….. ….. …..

#17.11.4. To compute the probability of a Big Bang coming into existence it is first necessary to calculate how many strings it would take to make a Universe which has a mass of 10^{53} grams (**section 17.7**).

The lightest elementary particle made of quantum strings is the electron neutrino which has a mass of less than 10^{-6} eV/c^2.
 According to **section 12:**   1 eV/ c^2 = 1.78 x 10^{-33} grams.
Thus an electron neutrino has a mass of less than 1.78 x 10^{-39} grams.

Since, no experimental evidence of strings has been found yet, a string must be much smaller than a neutrino and difficult to detect. Perhaps, a string is about 100 times smaller than a neutrino, so it would take around 100 strings to make a neutrino. A string would then have a mass of about 10^{-41} grams

Thus it would take about 10^{53} grams/10^{-41} grams/string = 10^{94} strings arranged exactly as shown in paragraph 3 above to create a Big Bang and a subsequent Universe (**sections 6.2 through sections 6.2.3**).

#17.11.5. The next step is to calculate the probability of these 10^{94} strings:

- BEING CREATED AT RANDOM,
- AT EXACTLY THE SAME PLACE IN EMPTY SPACE BEYOND OUR UNIVERSE,
 - IN EXACTLY THE SEQUENCE IN 3 ABOVE.
The probability result below would be the same if the sequence in paragraph 3 is one-dimensional, 2-dimensional (as shown), 3-dimensional, ..., n-dimensional.

Starting with string "R" the probability of the next string being correct "W" is:
$$1 \text{ out of } 4 = 1/4$$
The probability of the next string "G" being correct is also 1/4.
 The probability P of the first three strings being correct is:

$$P = 1(1/4)(1/4) = (1/4)^2 = (1/4)^{n-1}$$

where n is the total number of different types of strings.

To make a Big Bang, the recursion formula is for the probability that enough strings to make a Big Bang will be in the same place in Empty Space at the same time is.:

$$P = (1/4)^{n-1} \quad \text{where } (n-1) = (10^{94} - 1)$$

#17.11.6. There are some simplifying assumptions as this activity is going on everywhere throughout the infinity of empty space. The interest here is to calculate the probability of a Universe being created near out Universe within 50 billion years or so of the Creation of our Universe and near enough to perhaps travel to in the future.

#17.11.7.   Following this line of reasoning and noting 83 x3 = 249, leads to an estimate that it takes 10^{249} years on average for a Universe to be created within 30 billion light years of our Universe.

The conclusion is: THERE IS VERY LITTLE CHANCE THAT A UNIVERSE HAS BEEN OR WILL EVER BE CREATED ANYWHERE NEAR OUR UNIVERSE WITHIN THOUSANDS OF BILLIONS OF YEARS OF THE CREATION OF OUR UNIVERSE.

#17.11.8. Making the assumption that 1 billion (10^9) virtual strings spontaneously arrive at every Location L in our Universe and Space beyond our Universe, the number of strings N at point L in a year is calculated as follows:

Number of strings N that arrive at every location L per year =

$N = 10^9$ strings/sec =
   (10^9 strings)/sec)(60 sec/min)(60 min/hour)(24 hours/day)(365.25 days/year) =
   31,557,600 x 10^9 strings/year =
$N = 32$ x 10^{15} strings/year

Since it takes (according paragraph 4 above 10^{83} strings to make a Universe, if everything went perfectly, it would take (10^{83} strings) /(32 x 10^{15} strings/year) = 3.17 x 10^{66} years for a universe to be created randomly any specific location in the in space beyond an existing universe (that's 3 followed by 66 zeros years).

If the above analysis describes the creation of a universe, it is exceedingly improbable that one universe will ever be created near enough and timely enough to be traveled to from another universe. (BUT INFINITY IS A LONG, LONG TIME AND ANYTHING THAT CAN HAPPEN WILL HAPPEN!)

In about 10^{75} years, the last Black Hole from the last dead star will have radiated away all its Hawking Radiation and our Universe will have completely disappeared. But not to worry, in another 10^{200} years, more or less, another Universe will spring up (or will have sprung up) somewhere nearby with a Big Bang.

#17.12. BAD NEWS FOR TRAVEL BEYOND OUR UNIVERSE.

 If you are a curious and adventurous reader such as the author, one of your goals might be to get in a fast spaceship and travel beyond our Universe. I have always thought that empty space there (beyond our Universe) would be much like empty space between our galaxies, except perhaps COLDER AND EMPTIER. Some physicists believe that there is neither space nor time beyond our Universe. One of my goals is to find out what's there for sure! Perhaps there is another Universe close by this one as well. I would like to find out while I am still able to travel.

It seems that all that I need to travel beyond our Universe is a hyper-speed space ship that travels at about 2 billion light years per hour. Then I can follow my dream. Three hours later, I would be beyond the Cosmic Background Microwave Radiation (CMBR), out of our Universe, and traveling beyond our Universe.

Thanks to discoveries by Edwin Hubble, and many others, we know there are more than 400 billion galaxies in our Universe perhaps as many as 800 billion. Each galaxy has about 3 trillion stars maybe as many a 800 billion. Each star has several planets, moons, etc. like our galaxy, the Milky Way.
Our Universe is expanding rapidly. As pointed out earlier, Galaxies do not expand but are rapidly moving away from each other; the space between galaxies is expanding). The further away a galaxy is, the faster it is moving away from other galaxies and "creating" space between the galaxies.

HERE WAS MY PLAN: I will find a galaxy right on the edge of the Universe, get in my spacecraft, travel to it , and then launch my spacecraft to transit the short distance to journey beyond our Universe.

I BEGAN CHECKING GALAXIES. First I determined that Our Planet Earth and Our Sun are at the center of our Universe. The Universe looks the same from Earth in all directions. Then I found our galaxy, the Milky Way, was also at the center of the Universe. No matter, I expected that as Our Planet, Sun, and Galaxy were the "most important in the Universe" they would be at the center.

As I checked all the other galaxies in the Universe, I quickly found no matter which way I looked, no matter which galaxy I checked, no matter how far away the galaxy, or what direction, EVERY GALAXY IN THE UNIVERSE WAS AT THE CENTER OF THE UNIVERSE. This observation verifies the Big Bang Theory of Creation of Our Universe.

Unfortunately, after a few days of contemplation, I concluded that my plan to escape from our Universe was fatally flawed. NO MATTER WHICH DIRECTION THAT I TRAVELED IN, NOR HOW FAR, I WOULD STILL BE AND ALWAYS WILL BE AT THE CENTER OF OUR UNIVERSE.

#17.13. OUR UNIVERSE WILL CONTINUE TO EXPAND DUE TO DARK ENERGY.

Galaxies (held together by a black hole) will continue move to farther apart but each will always be at the center of our Universe. Ultimately, each galaxy will be alone, its stars will burn out, and its Black Holes will dissipate due to Hawking Radiation. Our galaxy, all other galaxies, and our Universe will entirely disappear. All that will be left is the empty space created as the galaxies moved apart from each other. NEVERTHELESS, ALL THE EMPTY SPACE THAT IS LEFT WILL BE SEETHING WITH VIRTUAL PARTICLES. SPACE AND TIME THERE WOULD BE JUST LIKE SPACE AND TIME IN OUR CURRENT UNIVERSE. IT WILL BE A GREAT PLACE FOR A NEW UNIVERSE TO BE BORN.

THIS METHOD OF THE BIRTH AND DEATH OF OUR UNIVERSE AND CREATION OF "EMPTY" SPACE DESCRIBED IN SECTIONS 17 THROUGH 17.13, LIKELY VERIFIES THE PREMISE THAT "ALL UNIVERSES EVERYWHERE ARE CREATED UNDER THE SAME CONDITIONS, IN THE SAME KIND OF "EMPTY SPACE" ENVIRONMENT, AS OUR UNIVERSE, AND THUS WILL BE SIMILAR TO OUR UNIVERSE IN ALL RESPECTS."

Now there is even a more exciting Adventure: Are there other Universes and are they just like ours as the data above seems to substantiate, or there 10^{500} possible kinds of Universes each with different physics and physics laws as some physicists believe? That's the ultimate challenge and the ultimate search. It just makes me want to jump in my spaceship and roar off at near the speed of light for a few years measured by my onboard clock, return to Earth when Earth clocks have advanced 10,000 years or so, and see how it all turned out.

#18. LONG BEFORE AND LONG AFTER THE BIG BANG: INFINITE EMPTY SPACE

What were things like before the Big Bang? Was everything everywhere migrating to be included in a small, ready-to explode-sphere of mass and energy the size of a marble, a proton, or a Planck length that unfurled to become our Universe? Was there nothing ever outside this sphere prior to the Big Bang – no nothing, no time?

Or more likely: There was infinite empty Spacetime just like empty Spacetime in our Universe except much, much colder ready to have our Universe born from a small sphere within it. There was infinite empty Space with other Universes elsewhere already born within it playing out their lives or having played out their lives from a Big Bang to an accelerated inflationary expansion into cold, cold dissipation throughout empty, infinite Spacetime. And there was infinite empty spacetime with other Universes yet to be born within it, with each Universe yet to play out its life. This seems (arguably) to be the simplest explanation. Assuming it is the correct explanation of this infinite empty Spacetime, **section 12.7.2** describes its characteristics.

When will our Universe come to an end? Our Universe will continue to expand at an accelerating rate driven by dark energy (**section 14.2.2).** Galaxies are rushing away from each other. Galaxies are not expanding at this time. The galaxies seem to have sufficient mass and gravitational force attraction to hold a galaxy together.

Meanwhile, stars are using up all their hydrogen fuel (that has been sustaining nuclear fusion). Stars are then expanding and dying or if large enough, becoming novas and supernovas, neutron stars, or ultimately black holes. Finally the black holes radiate all their energy away down to the last photon paying back the energy borrowed from empty spacetime in full. Finally, nothing is left except **NEGATIVE-ENERGY** empty spacetime with the cycle going on again – universes forming and universes dying.

Fortunately, our Universe has many billions of years of life left. After all it is only 13.8 billion years old; there still is plenty of time for our descendents to find a way to jump to a younger and hopefully hospitable universe.

#19. UNDERSTANDING NATURE, OUR UNIVERSE, AND BEYOND OUR UNIVERSE USING SCIENTIFIC METHOD.

The approach in **sections 6, 17, and 18** to explain the origin and beyond our Universe is probabilistic in nature. Einstein reportedly said to the effect about quantum theory, that nature was not probabilistic and did not throw dice. He later changed his mind as the evidence of a Universe based on probabilities became overwhelming. (See **section 12.4.**)

The material herein is about particle physics, relativity, and cosmology. These subjects are about understanding nature and the Universe using the scientific method. The scientific method explains the workings of nature by forming hypotheses (conjectures) and theories about exactly how nature works. In physics, there is no science without mathematics. In physics, there is no science without repeatable experiments to verify or disapprove the hypotheses, theories, and mathematics. Physics rises or falls on the results of experiments. There is no science without experiment. Until experimental results are in, the hypotheses, theories, and mathematics are all just a bunch of (hopefully educated) conjecture. Nevertheless, particle physicists have been stunningly successful to date in having experiments ultimately confirm many of their theories and much of their mathematics.

In particle physics and cosmology, there was a lot riding on finding the Higgs boson that was predicted to exist 48 years ago by Peter Higgs and his colleagues François Englert, Gerald Guralnik, Carl Hagen, and Tom Kibble. The Higgs boson was finally found in July 2012. Finding the Higgs boson confirmed the theory and mathematics of the Standard Model (SM) of particle physics. Particle physicists everywhere breathed a sigh of relief.

Now particle physicists eagerly await the results of new experiments at CERN that will enable them to push forward to verify their theories of their predicted Graviton, Super-Symmetric Particles, Strings, Membranes, 11 Dimensional Spacetime, and Universe Creation. Had the Higgs boson not been found, some physicists would have said, "I told you so!" and perhaps won a bet. But no matter what the results at CERN, all physicists will go back to eagerly revise their theories and mathematics as necessary to bring them into agreement with the experimental results, the findings of the latest hard science, the repeatable experimental evidence, and repeatable results and conclusions. Then they will rush to extend the findings to new frontiers.

For example, the Higgs boson was not considered to be found until the extensive repeatable experimental evidence at CERN was valid to 99.999 percent. Prudent observers should consider carefully done repeatable experiments a revelation no matter what field the scientific experiments are in.

By applying the scientific method, science seems to be progressing nicely by leaps and bounds! Civilization on the other hand seems to have not progressed from the same ages-old problems. If our civilization survives (or doesn't survive) the current religious wars, drug wars, ethnic wars, other wars, discrimination of every type, political standoffs, pollution, water shortages, energy shortages, slavery, starvation, poverty, climate changes, crime, and other evils, physicists and other scientists will continue to go where the mathematics and experimental results take them. Stephen Hawking states in his referent video that he attended a 1985 science meeting in Rome with Pope John Paul II. According to Hawking, Pope John Paul cautioned the attendees that it was alright to explore how the Universe works but not its origin. Hawking in his video responds, "I am glad to say that I for one haven't followed his (Pope John Paul's) advice. I can't simply switch off my curiosity."

Besides predicting as yet unseen super-symmetric particle for every elementary particle, some of the mathematical physics currently (without any direct experimental confirmation) predicts the origin of our Universe, multi-universes, bubble universes, parallel universes, worm holes, and the like. The validity of the theories and conjectures rests on mathematics and experiments that, in the case of super-symmetric particles, are now going on at CERN. In other cases, the relevant experiments may not be performed or performable for dozens or even hundreds of years, if ever. The physics will go ultimately where the experiments lead, and there will be many surprises.

The question to ask physicists (and other scientists) is: "How will their conjectures and theories of a Universe be tested and proven or disproven by repeatable experiments?" That is the question physicists and scientists ask themselves every day!

#PART IV. OVERVIEW OF MATHEMATICS OF QUANTUM THEORY AND RELATIVITY

So far, this work has concentrated on the physical descriptions of the world of quantum and particle physics with only limited emphasis of the mathematics involved.
This policy will be continued. However, it would be unjust not to include just a few examples for the reader to take note of the elegance and precision of the mathematics that, long in advance in many cases, theorized the counter-intuitive aspects of our Universe at its extremes. And, when experiments finally confirmed the mathematical predictions, the accuracy of the mathematics and experimental confirmation in many instances was to more than 10 significant figures. This accuracy is like firing a bullet from New York to London and hitting a 1 cm target.

#20. SCHRÖDINGER'S EQUATION

#20.1. PROBABILISTIC CHARACTERISTICS OF THE WAVE-PARTICLE QUANTUM WORLD.

The previous **section**s have described the Universe as being composed of objects that have characteristics of both particles and waves. For large objects, only particle characteristics are important as defined by the kinematics of Isaac Newton and Albert Einstein. For small objects of the quantum, particularly for elementary particles, wave characteristics are of major importance.

In 1925 Edwin Schrödinger was among the first to adopt Newtonian mechanics of large objects and the mathematics of wave motion (for a vibrating string, sound, and the like) to quantum mechanics of small objects. For this, he received the Nobel Prize in 1936.

Schrödinger defined a wave function (wave amplitude) ψ for a quantum particle.
The quantity ψ is a function of (depends on) space coordinates x, y, z and time t. It has several properties:
- It can interfere with itself and exhibits diffraction characteristics.
- It is large where the particle has a large probability of being located.
- It exhibits the behavior of a single quantum particle such as a photon or electron not a number of particles.
- Its square ψ^2 mathematically expresses the probability of finding the quantum particle at any point x, y, z. The sum of all the probable locations is equal to one (1.00).
- It does not ordinarily interfere with other quanta. But, as will be shown later, the original formulation of Schrödinger's equations will be expanded so some quanta (specifically, strong and weak force bosons) can interact with each other.

#20.2. CONSERVATION CHARACTERISTICS OF QUANTUM WORLD.

In any interaction, theoretical and experimental physicists consider all the various kinds and forms of the conserved quantities involved. For example, the total energy of any interaction is conserved: the same before and after an interaction including all the work done. Energy may be changed in form: kinetic to potential, potential to kinetic, mass to energy, energy to mass, chemical to heat, mechanical to electrical, and so forth. Momentum is another key conserved quantity. So is probability. So is angular momentum.

Perhaps the most important conservation principle is the conservation of energy. In quantum mechanics, this is expressed as Schrödinger's wave equation. It is derived as follows:

$$E_{Total} = E_{Kinetic} + E_{Potential} = E = K + V$$

The classical (Newtonian) relation for Kinetic Energy K is:
$$E_{Kinetic} = K = \tfrac{1}{2}mv^2$$

Non-quantum kinetic energy is expressed in terms of mass m, velocity v, and momentum p as:

$$p = mv,$$

$$E_{Kinetic} -= K = \tfrac{1}{2}mv^2 = p^2/2m$$

de Broglie's equation is expressed in terms of wavelength λ, Planck's constant h, frequency v, and the speed of light c.

$$\lambda = h/p \qquad \lambda = c/v \qquad v = c/\lambda \qquad p = h/\lambda = 2\pi\,\bar{h}\,/\lambda$$

As Planck pointed out, each discrete quantum particle contains an energy E.

$E = hv = hc/\lambda = 2\pi \overline{h}\, v$

$p = 2\pi \overline{h}\, /\lambda$

$\overline{h} = h/\, 2\pi \qquad h = 2\pi\, \overline{h}$

Where h is Planck's constant.

The wave equation for a wave function ψ of a free particle depends on **vector** position **r** as a function of (**x, y, z**) or $(r, \theta, \emptyset)$ and time t. The simplest wave equation for a vibrating string or sound waves for instance is of the form:

[1] $\qquad \dfrac{\partial 2}{\partial t2}\, \psi \;=\; k\, \dfrac{\partial 2}{\partial r2}\, \psi \;=\; k\, \nabla^2 \psi$ \qquad\qquad Where k is a constant.

$\dfrac{\partial 2}{\partial r2}\psi$ has the meaning of $\nabla^2 \psi$ and is the rate of change of velocity of a particle with respect to Its coordinates:

$$\nabla^2 \psi = \left(\dfrac{\partial 2}{\partial x2} + \dfrac{\partial 2}{\partial y2} + \dfrac{\partial 2}{\partial z2}\right)\psi$$

Of several solutions to this equation, the k that is most applicable to quantum particles is:

$k = i\overline{h}\, E/p^2 \;=\; i\overline{h}\, /2m$

$i\overline{h}\, E = \; i\overline{h}\, p^2/2m$

Substituting into [1] above:

$\dfrac{\partial 2}{\partial t2}\, \psi = i\overline{h}\, /2m\; \nabla^2 \psi$

20.2. CONSERVATION CHARACTERISTICS OF QUANTUM WORLD (CONTINUED).

Multiplying through by

$$i\overline{h} \; \frac{\partial 2}{\partial t2} \; \psi = (i\overline{h}) \, (i\overline{h} /2m) \; \nabla^2\psi = - \overline{h}^2/2m) \; \nabla^2\psi$$

For a quantum particle, the energy relation for total energy E = K + V for at any time t can be expressed in terms of its wave function ψ as follows:

$$E \rightarrow i\overline{h} \; \frac{\partial}{\partial t} \, \psi$$

The kinetic energy K can likewise be expressed for a quantum particle at any **vector** position **r** can be expressed in terms of its wave function ψ as follows:

$$K \rightarrow - i\overline{h}^2/2m \; \frac{\partial 2}{\partial r2} \, \psi = - i\overline{h}^2/2m \; \nabla^2\psi$$

Therefore: the equation for the energy E or a quantum particle (wave) and its kinetic energy K is:

$$i\overline{h} \; \frac{\partial}{\partial t} \, \psi = - i\overline{h}^2/2m \; \nabla^2\psi$$

The above equation can now be generalized to the Schrödinger equation for conservation of energy for any quantum particle by including the potential energy V:

$$E_{Total} = E_{Kinetic} + E_{Potential} = E = K + V$$

$$i\overline{h} \; \frac{\partial}{\partial t} \, \psi = - i\overline{h}^2/2m \; \frac{\partial 2}{\partial r2} \, \psi + V(\mathbf{r}, t) \, \psi$$

This is the Schrödinger wave equation for the conservation of energy in the motion of a free quantum particle. Potential energy V(**r**, t) is a function of **r** and t.

#20.3 EXPANDING SCHRÖDINGER'S EQUATION TO RELATIVISTIC PARTICLE INTERACTIONS.

Additions are necessary to Schrödinger's equation to describe how one quantum particle interacts with other particles through mediation of force bosons. The force bosons are the electromagnetic force photon Υ , eight strong color force gluons g, weak force bosons W^+, W^-, and Z^0, and the gravitational force bosons – the Higgs boson(s) and the as yet undetected graviton. These additions were not easy for theoretical physicists to accomplish. Many turns and twists occurred that required tremendous intellect and imagination to work around. Many, many of the additions tried by the theorists caused equations to blow up with infinities and inconsistencies.

 Several the books in the **BIBLIOGRAPHY** tell the magnificent stories of the participants and their wonderful achievements and include the detailed mathematics involved. Their efforts culminated in the prediction of the Higgs boson and its experimental detection in June 2012. The story is not yet complete with additional Higgs bosons predicted but not yet detected and the graviton not yet detected, as well as many advanced topics under intensive investigation as described in **Part III** herein.

#21. SPECIAL RELATIVITY.

#21.1. OVERVIEW OF SPECIAL AND GENERAL RELATIVITY - ENERGY WARPS SPACETIME.

The mathematics of calculus and physics of motion developed by Isaac Newton will always remain at the foundation of physics. However, as will continue to occur in science as understanding of and experimentation about our universe continues, expansions and add-ons to the work of earlier Giants will be necessary. One of these add-ons, concerned with electricity and magnetism, was Maxwell's equations (1873). The mathematical inconsistency in separate equations for electrical and magnetic fields formulated by earlier investigators was resolved by Maxwell.

Maxwell's equations also had the unexpected result of predicting that all electromagnetic waves including light propagate at the same constant velocity of the speed of light for all observers regardless of the relative velocity of the source to the observers and the constant relative velocity of the observers to each other.

This finding of the constant velocity of the speed of light by Maxwell was experimentally confirmed by the famous experiment of Michelson and Morley in 1887 for which a Nobel Prize was awarded in 1907. Measurements were made of light velocity by observers as the Earth rotated around at about 1609 kilometers (1000 miles) per hour at the equator toward light from the Sun and away from light from the Sun. This is at variance with other waves (**section 3**). **Physicists were astounded by the findings**. In 1905 Einstein's explanation was provided in his Special Theory of Relativity.

Several excellent books are on the mathematics of Special and General relativity are listed in the Bibliography ranging from elementary [**Stannard**] to comprehensive [**Collier**].

#21.1.1. KINETIC ENERGY WARPS SPACE AND TIME (SPACETIME) COLLINEARLY.

The remarkable finding by Maxwell of the constant velocity of the velocity of light was one of the key developments that led Einstein to his Special Theory of Relativity **(Section 21)** and later to his General Theory of Relativity **(section 22)**. In his Special Theory, Einstein postulated that the speed of light c **(section 3.5)** is measured to be the same regardless of the:

- Relative velocity of observers in inertial (constant velocity) frames moving with constant velocity toward or away from each other and the source of light.
- Constant velocity of the light source moving toward or away from observers in inertial frames of constant relative velocity.

Einstein concluded that space and time are intertwined into an entity spacetime – a term that is used throughout this book. The amount of this warpage is dependent on the kinetic and potential energy of an object. The amount of spacetime warpage caused by kinetic energy of an object determines the amount of spacetime warpage and the shortest straight-line (collinear) path a free-falling object follows in spacetime.

The effect of this spacetime warpage by the kinetic energy (movement) of an object causes clocks on the moving object to be measured to run slower than clocks on a stationary object (time dilation).

Further, the spacetime warpage causes an observer on one inertial frame to measure distances on other inertial frames to be less than measured by observers on the other frames (distance contraction).

The spacetime where Special Relativity applies is "flat spacetime." Flat spacetime is free of the effects of gravity as gravity causes spacetime to curve. Curved spacetime is covered in section 22 (General Relativity). Nevertheless, there are many places where no large objects are close, the effects of gravity are minimal, and Special Relativity is applicable.

If the kinetic energy (velocity) of an object increases, its clocks run slower and slower as compared to clocks on stationary objects and approach stopping as its velocity approaches infinity. Also, as the velocity of the moving object approaches infinity, the distances measured by the moving object as compared to distances measured by stationary objects approaches zero.

Thus, an object of infinite velocity travels to infinity in zero seconds!! But of course, no object possessing mass can ever attain an infinite velocity (section 22.1). It would take infinite energy to accelerate an object that had any mass to infinite velocity.

Examples of Special Relativity in nature:

- Massless probability waves (**section 12.4.2**) are thought to exist and travel out in all directions at infinite velocity. Probability waves may one day provide a means of instant communication anywhere in the Universe.

- Muons (**section 9**) are "heavy" electrons that are created in the upper atmosphere by collisions of powerful cosmic rays from the sun with protons. Muon lifetime is less than a millionth of a second. Yet muons are found to exist much longer and travel much further than expected. The explanation for this is that a muon's clock runs much slower when measured by a stationary observer's clock on the Earth. A muon's clock only gets up to a millionth of a second before the muon decays and ceases to exist, but a stationary clock on Earth indicates it lives much longer. Further, due to their high velocity, the distance the muon measures that it travels is much shorter than the distance measured by an observer on the Earth.

#21.1.2. POTENTIAL ENERGY WARPS (CURVES) SPACE AND TIME (SPACETIME).

After finishing his Special Theory of Relativity in 1905 covering the effect of kinetic energy on spacetime, Einstein turned to the effect of potential energy on spacetime. It took him another 10 years to complete his General Theory of Relativity which describes how potential energy curves paths of free-falling objects in spacetime. **Section 22** describes General Relativity.

#21.1.3. BEHAVIOR OF LIGHT IS DIFFERENT THAN THAT OF OTHER TYPES OF WAVES.

Section 3.0 explains the differences between light waves and other types of waves. For other waves, the relative velocity of the source and observers must be added or subtracted.

Postulates about the velocity of light and its associated physics are:

- The laws of nature are the same for all inertial (constant velocity) frames of reference.
- An object at rest in one inertial frame is either at rest or moving at a constant velocity with respect to all other inertial frames.
- Objects in motion in an inertial frame keep moving in the same direction at the same velocity unless acted upon by a force.
- Any frame moving at a constant velocity with respect to an inertial frame is an inertial frame.

As long as the distances are short and uniform (constant) velocity differences between observers are small, events seem to observers to occur at exactly the same time even if in different inertial (constant velocity) frames.   The relative velocities (v1, v2, ...) of the frames are merely just added or subtracted depending upon whether approaching or moving away from each other.

However, as the relative velocity between different observers in various inertial frames increases and nears the speed of light, all observers in inertial frames regardless of their relative velocities and the velocity of the source of the light find the speed of light is measured to be the same.

In addition, distances and time of a specific event, as measured by various observers in various inertial frames, will be different giving seeming bizarre results including being able to travel ahead in time (but not backward).

The coordinate equations of Lorentz (and others) to transform the distances and time of an event in one inertial frame to another were well known before Einstein's efforts. It took an Einstein in his Special Theory of Relativity, however, to give a qualitative and physical explanations as well as a mathematical explanation of what was happening.

One question that may arise to the reader is, "Why does light have these astounding constant velocity characteristics?"

> - Light is electromagnetic energy. Electromagnetic energy is all pervasive in our Universe from all the charged elementary and composite particles listed in **sections 9 and 10** to atoms to elements, to compounds, to planets, to stars, to black holes, and to galaxies. None of these would exist unless light and electromagnetic energy exhibited strange relativistic behavior on this all pervasive grand scale that our day-to-day experiences do not prepare us for.

Of course, Nature is full of such surprises and bizarre behavior. Many, many are described herein such as water decreases in volume as it cools until below 4 degrees when it expands so ice floats rather than sinks, objects exhibit both wave and particle characteristics, and so forth. But who would suspect the light which is visible and in which we are completely immersed everyday would hold such surprises?

Some Physicists believe (mathematically) that there are untold trillions of different universes that could exist with untold trillions and trillions of different laws of physics and bizarre behaviors. It seems much more likely (to the Writer) that in order to exist, all other Universes must be created with and obey the same Laws of Physics just like our Universe and all others will fizzle out in just moments after their creation.

#21.1.4. ELECTROMAGNETIC WAVES PROPAGATE FROM CENTRAL POINT.

In physics, consistencies and inconsistencies in the mathematical formulation many times lead to unforeseen insights. Such was the case with Maxwell's theory of electromagnetic wave propagation and, as it turned out with Einstein's Special and General Relativity as well:

The following two equations describe a spherical electromagnetic light wave-front being propagated at the speed of light (c) from a central point along the axis (x, y, z) at time (t) for two observers in different inertial frames: a stationary observer x (unprimed) and a moving observer x' (primed). This is similar to the path Einstein followed rather than using the Lorentz equations which are described below.

$$x^2 + y^2 + z^2 = (ct)^2 \qquad \text{(Unprimed Observer)}$$
$$x'^2 + y'^2 + z'^2 = (ct')^2 \qquad \text{(Primed Observer)}$$

Or if motion is aligned only with the x axis or x' axis,

$$x^2 = (ct)^2$$
and, $$x'^2 = (ct')^2$$

The laws of physics in inertial frames are the same for both the stationary and moving observers. This provides a means to translate from the coordinate system of a stationary observer (unprimed) to that of an observer (primed) moving at a constant velocity v with respect to the stationary observer:

$$x^2 + y^2 + z^2 - (ct)^2 = x'^2 + y'^2 + z'^2 - (ct')^2$$

If the unprimed and primed axes are aligned and motion is along only the x axis, then:

y = y' and z = z'.
$$x^2 - (ct)^2 = x'^2 - (ct')^2$$

Solving this last equation leads to the familiar Lorentz equations below in **section 21.2**. The converse showing that the Lorentz equations lead to the above equations for the propagation of light is shown in **section 21.4.**

#21.2. LORENTZ EQUATIONS.

Assume there are constant velocity inertial systems such as trains (or busses, or airplanes, etc), all moving parallel at different constant velocities. You are on one. Since you are moving at constant velocity, your train appears to you to be standing still. Other observers on their different trains appear to themselves to be standing still. Each observer on all the different trains (inertial platforms) believes they are standing still, but all the other observers on other trains are moving (at constant velocity).

Select any other train moving with respect to you at constant velocity v. Both your train and the other train have identical light clocks that emit a pulses of light that travel to a receiver on the ceiling a distance d away. As the two trains pass, the identical clocks are synchronized. In the below discussion, the following equation for distance applies:
        (distance d) = (velocity v) times (time t).   (d = vt)
Later the distance equation for x (distance) will be used. (x = vt).

As you watch the clock on the other train, you realize that its light pulse has further distance d_m to travel taking time t to reach its ceiling receiver than the pulse from your clock to reach its ceiling receiver due to the movement of the moving train with respect to your "stationary" train. While the light from the moving clock is in route from the source to the receiver, the moving train has traveled a horizontal distance d_h

Distance d_m for one pulse of moving $d_m = ct$
clock according to a stationary observer

Distance d' for one pulse of moving clock $d' = ct'$
on moving other train according to
moving observer

Horizontal distance d_h moving train has traveled in one pulse: $d_h = vt$

The additional distance the light pulse must travel is dependent on the distance (vt) the other train travels in the time t it takes for the light pulse to reach the ceiling of the moving train as seen by an observer in stationary train.

The above distances make a right triangle that is solved by the Pythagorean Theorem:

$$(ct)^2 = (ct')^2 + (vt)^2 \qquad\qquad c^2t^2 = c^2t'^2 + v^2t^2$$

$$c^2t'^2 = c^2t^2 - v^2t^2 = (c^2 - v^2)t^2$$

$$t'^2 = (1 - v^2/c^2)t^2$$

$$t' = t(1 - v^2/c^2)^{1/2} = \Upsilon t \qquad \text{where } \Upsilon = (1 - v^2/c^2)^{-1/2}$$

.

The above equations lead immediately to the Lorentz equations for relative distance and time, where

     x = (v)(t): distance x in stationary inertial frame,

     t = time in stationary inertial frame

     x' and t' are distance and time in the inertial frame moving at constant velocity v.

$$x' = (x - vt)[1 - (v/c)^2]^{-1/2} \qquad x = (x' + vt)[1 - (v/c)^2]^{-1/2}$$

$$t' = [t - (vx/c^2)][1 - (v/c)^2]^{-1/2} \qquad t = [t' + (vx'/c^2)][1 - (v/c)^2]^{-1/2}$$

$$x' = \Upsilon(x - vt) \qquad\qquad x = \Upsilon(x' + vt)$$

And, $\qquad t' = \Upsilon(t - vx/c^2) \qquad\qquad t = \Upsilon[t' + (vx'/c^2)]$

Note that the equations for the primed (x', t') and unprimed observers (x, t) are identical except for the sign of velocity v. Both the primed and unprimed observers believe that they are the stationary observer and the other is moving at a constant relative velocity (v). It is impossible for either observer to determine otherwise.

#21.3. LENGTH CONTRACTION AND TIME DILATION.

The following statements summarize Special Relativity:

- Observers on different inertial frames moving at different constant relative velocities all believe that their frame is stationary.

- A stationary observer will measure all moving clocks to run slower.

- Moving observers will measure distances in stationary frames to be less than measured by the stationary observers.

- **Stationary observers and moving observers** all think they are a **stationary observer**! No matter what the relative (constant) velocities of various frames, all clocks of all observers (that they always have with them) just keep ticking merrily along. However, stationary observers will measure all moving clocks to run slower!

No matter what, life cannot be prolonged for a moving observer. A stationary observer may think the moving observer lives hundreds or thousands of years longer than they (the stationary observer) do as their moving clock is running slower than theirs, but the moving observer won't notice – their clock just keeps ticking normally. Life may be prolonged on a moving clock as measured on a stationary clock, but the moving observer won't notice other than perhaps the stationary observer may have long since expired.

Time is not dilated; distances are not contracted for the moving observer on the moving frame or the stationary observer on the stationary frame. However, and this is the whole point, the ticking of clocks on moving frames will be measured to run slower by the stationary clock, and the distances measured on the stationary frame by the moving observer will always be contracted - less than the distance measured than by the stationary frame by the stationary observer.

#21.3.1. LENGTH CONTRACTION.

For length $(L = x_2 - x_1)$ in all other frames, as seen by an observer in the moving frame:

$$x_2' - x_1' = (x_2 - vt)[1 - (v/c)^2]^{1/2} - (x_1 - vt)[1 - (v/c)^2]^{1/2}$$

$$x_2' - x_1' = L' = (x_2 - x_1)[1 - (v/c)^2]^{1/2} = L[1 - (v/c)^2]^{1/2} = YL$$

Length (L), of a object in the stationary observer's frame, will appear to be contracted (less) to the moving observer by the factor $Y = [1 - (v/c)^2]^{1/2}$ which is always less than one. This is known as length or distance contraction. The moving observer will measure distances IN THE STATIONARY FRAME to be less than measured by the stationary observer. This contraction is only along the axis of motion; in this case, only along the x and x' axes.

#21.3.2. TIME DILATION.

$$t'_2 - t'_1 = T' = (t_2 - t_1)\,[1 - (v/c)^2\,]^{-1/2} = \Upsilon T$$

When the stationary observer's clock has moved one hour, the stationary observer will find the MOVING CLOCK has moved less than one hour. The stationary observer will find the moving clock is running slow.

An observer's own frame is always stationary, but is seen by observers in other frames (of different constant velocity) to be moving, who in turn each believe that their frame is stationary and the others are moving

Time in an observer's own frame as indicated on the observer's own clock just keeps on ticking at its regular rate and never runs slow or fast. TIME IN ALL OTHER FRAMES MOVING WITH RESPECT TO THE STATIONARY OBSERVER is always dilated (slowed down) as measured AND COMPARED TO TIME MEASURED by an observer in their stationary frame.

As a moving observer's velocity increases toward the speed of light, with respect to a "stationary frame" the moving observer will measure distances in the stationary frame to decrease. Time in the moving frame as measured time by clocks in the stationary frame will slow. As relative velocity approaches the speed of light, distances in the stationary frame will approach zero and time on the moving clock will approach zero.r.

However, so long as the relative velocity of the primed (moving) and unprimed (stationary) observers remains unchanged, all observers believe that they are the stationary observer and the other observers are the moving observer and there is no way for any to determine otherwise. All observers will continue to come to the same conclusions about distance contraction and time dilation in the other's frames unless the constant velocity relationship of the observers is disturbed by one accelerating (or decelerating) as described in the twin paradox (**section 21.12**).

For example, a stationary observer might measure a distant star to be a distance x of 20 light-years away. A space traveler starting at the location of the stationary observer and moving at a relative constant velocity (v) of 0.6c toward this star would find it to be a distance L' of only 16 light-years away.

$$L' = L \, [1 - (v/c)^2]^{1/2} = 20 \, [1 - (0.6c/c)^2 \,]^{1/2} = 16 \text{ light-years}$$

The stationary observer's clock (t) would find the space traveler's moving clock (t') runs slow:

$$T = T' \, [1 - (v/c)^2]^{-1/2} = 1.25 \, T'$$

Or, for every hour the space traveler's moving clock time (T') advances, the stationary observer would find the stationary clock time (T) advances 1.25 hours and the moving clock is deemed by the stationary observer to run slower. If the space traveler comes to a screeching halt (decelerates) at the distant star and turns around and returns to earth (accelerates), then the stationary observer would find the space traveler will indeed have aged less than those left behind on Earth (the twin paradox).

#21.4. LORENTZ EQUATIONS LEAD TO THE EQUATIONS FOR THE PROPAGATION OF LIGHT.

The initial formulation of Special Relativity equations based on the propagation of light at a constant velocity:

$$x^2 + y^2 + z^2 = (ct)^2 ,$$
$$x'^2 + y'^2 + z'^2 = (ct')^2$$

Also, $x^2 + y^2 + z^2 - (ct)^2 = x'^2 + y'^2 + z'^2 - (ct')^2$

And, if $ct = iT$ Where $\sqrt{-1}$ is the square root of -1. This substitution was recommended to Einstein by the mathematician Minkowski, to bring Einstein's notation into the notation of standard infinite-dimensional linear algebra.

$$x^2 + y^2 + z^2 + (T)^2 = x'^2 + y'^2 + z'^2 + (T')^2$$

If the unprimed and primed axes are aligned and motion is along only the x axis, then y = y' and z = z' then, the equations for the propagation of light become:

$$x^2 - (ct)^2 = x'^2 - (ct')^2$$

The Lorentz equations are:

$$x' = (x - vt)[1 - (v/c)^2]^{-1/2} \qquad x = (x' + vt)[1 - (v/c)^2]^{-1/2}$$

$$ct' = [ct - (vx/c)][1 - (v/c)^2]^{-1/2} \qquad ct = [ct' + (vx'/c)][1 - (v/c)^2]^{-1/2}$$

Substituting these Lorentz transforms into the above two equations x'^2 and $(ct')^2$ for the propagation of light for an observer moving at a relative velocity v:

$$(x')^2 - (ct')^2 = [(x)^2 - 2\,vtx + (vt)^2 - (ct)^2 + (2vtx) - (vx/c)^2][1 - (v/c)^2]^{-1}$$

$$(x')^2 - (ct')^2 = [(x)^2 - (ct)^2 + (vt)^2 - (vx/c)^2][1 - (v/c)^2]^{-1}$$

Now factor out: $(x^2) - (ct)^2$

$$(x')^2 - (ct')^2 = [(x)^2 - (ct)^2][1 - (v/c)^2][1 - (v/c)^2]^{-1}$$

Thus, $(x')^2 - (ct')^2 = (x)^2 - (ct)^2$

This interesting result shows that the Lorentz Transform leads directly to the propagation of light at a constant velocity regardless of the relative velocities of the reference frames. Lorentz (and others) working strictly from early experimental results such as obtained by Michelson and Morley, and Einstein working strictly from theoretical considerations, arrived at the same equations for translating the place and time of the occurrence of an event of two constant velocity observers from one coordinate system (unprimed) to another (primed) and vice versa.

#21.5. HYPERBOLIC FUNCTIONS OF LORENTZ EQUATIONS.

The Lorentz equations transform coordinates from x to x' and ct to ct' in hyperbolic space using hyperbolic functions. The Lorentz equations are:

$$x' = [(x) - (vt)] [1 - (v/c)^2]^{-1/2}$$

$$ct' = [(ct) - (vx/c)] [1 - (v/c)^2]^{-1/2}$$

Using the identity, $(\cosh^2 \Omega - \sinh^2 \Omega = 1)$, and these hyperbolic functions,

$$\cosh \Omega = [1 - (v/c)^2]^{-1/2}$$
$$\sinh \Omega = (v/c) \cosh \Omega = (v/c) [1 - (v/c)^2]^{-1/2}$$
$$\tanh \Omega = \sinh \Omega / \cosh \Omega = v/c$$

For a rotation in hyperbolic space, let:
$$x' = x \cosh \Omega - (ct) \sinh \Omega$$
And, $$ct' = - x \sinh \Omega + (ct) \cosh \Omega$$

Then, $x' = x \cosh \Omega - (ct) \sinh \Omega = x [1 - (v/c)^2]^{-1/2} - (ct) (v/c) [1 - (v/c)^2]^{-1/2}$
$x' = (x - vt) [1 - (v/c)^2]^{-1/2}$

$ct' = - x \sinh \Omega + ct \cosh \Omega = [-(vx/c) + (ct)] [1 - (v/c)^2]^{-1/2})$
$t' = (t - vx/c^2)) [1 - (v/c)^2]^{-1/2}$

And, $x^2 - (ct)^2 = x'^2 - (ct')^2$
$$= x'^2 \cosh^2 \Omega - 2ctx \sinh \Omega \cosh \Omega + (ct)^2 \sinh^2 \Omega$$
$$+ x^2 \sinh^2 \Omega + 2ctx \sinh \Omega \cosh \Omega - (ct)^2 \cosh^2 \Omega$$
$$= x'^2(\cosh^2 \Omega - \sinh^2 \Omega) + (ct)^2 (\sinh^2 \Omega - \cosh^2 \Omega)$$
Thus, $x^2 - (ct)^2 = x'^2 - (ct')^2$

The transformation of x and ct to x' and ct' using hyperbolic functions (the Lorentz equations) leads to rotation of coordinates with tanh Ω = v/c, the greater v, the greater the angle Ω of rotation.

Unfortunately, the rotation is in hyperbolic coordinate space rather than the rectilinear Cartesian coordinate space we live in.

This is perhaps more than most readers need or want to know about "hyperbolic" space. It was included because as this **section** and the next unfold, the mathematics will lead to various kinds of spaces, times, and coordinates for which the mathematics may not always represent reality but can be easily transformed into more relevant functions such as done in the next **section.**

#21.6. COORDINATE TRANSFORMATION USING LINEAR CARTESIAN COORDINATES.

Sections 21.4 and 21.5 noted that as the equations of Special Relativity as originally formulated must hold for all parameters and coordinates, we can write,

$$x^2 + y^2 + z^2 - (ct)^2 = x'^2 + y'^2 + z'^2 - (ct')^2$$

This leads to formulation in unwieldy hyperbolic space as described in **section 21.5**. The highly esteemed mathematician Minkowski with whom Einstein was working noted, that if the Lorentz equations expressed in hyperbolic space, were modified with a substitution of variable, they could be placed within the complete framework of advanced vector and tensor analysis and linear algebra:

$$a^2 + b^2 + c^2 + ... = S^2$$

Where S is the non-negative square root (absolute value) of S^2 is called the norm or length of S. This equation is the familiar Pythagorean Theorem generalized to an infinite number of orthogonal dimensions a, b, c, ..., termed "vector space. S is a vector and is the vector sum of vectors a, b, c, ... The norm is the vector length of the hypotenuse calculated using the Pythagorean Theorem.

Minkowski urged Einstein to introduce a new variable ct = iT , where i = $\sqrt{-1}$ (square root of -1). With this substitution, the theory of Special Relativity can be rewritten as,

$$x^2 + y^2 + z^2 + (T)^2 = x'^2 + y'^2 + z'^2 + (T')^2 = S^2$$

Now, all observers (moving and stationary) would agree on the vector-sum length (norm) of an event S and where and when S took place in space. The vector distance (x, y, z,) and "vector-spacetime" (T) coordinates at which it took place would be different for their specific relative velocities. Each observer could translate from their coordinate system to any other coordinate system and vice versa. Each would find a different vector distance to and vector-spacetime of a specific "event" S that happened such as arrival at a specific location in space. Yet all would agree that such an event happened. Also, all would agree on the vector values of each S1 and S2.

Two observers calculating the vector-spacetime interval between two "events" $S_2 - S_1$ each using the formula for their "stationary" reference frames would get the same answer for $S_2 - S_1$ even though their reference frames were moving at different constant velocities with all the other reference frames.

$$(S_2 - S_1) = (x_2 - x_1)^2 + (y_2 - y_1)^2 + (z_2 - z_1)^2 + (T_2 - T_1)^2$$

The x, y, z, and T would have different values for all different reference frames x', y', z', and T' but the value calculated by all for S1, S2, and $S_2 - S_1$ would be the same.

$$(x_2 - x_1)^2 + (y_2 - y_1)^2 + (z_2 - z_1)^2 + (T_2 - T_1)^2 = (S_2 - S_1)^2$$

$$= (x'_2 - x'_1)^2 + (y'_2 - y'_1)^2 + (z'_2 - z'_1)^2 + (T'_2 - T'_1)^2$$

- Different observers moving in different reference frames at different relative velocities would all agree that an Event S_1 such as launching a spacecraft took place.
- They would all agree that an Event S_2 took place such as the spacecraft arriving at a destination place and the vector parameters of S2.
- They would all agree S_1 happened before S_2.
- They would all agree on the difference $(S_2 - S_1)$ between the events.
- They would not agree on the specific vectors (x, y, z, T) where and when specific Events (S) took place.
- Four parameters (x, y, z, and T) are needed by each observer to precisely measure the occurrence of an event in "vector-spacetime" and these four are different for each different observer.

Einstein now also realized that he wanted to formulate his General Theory of Relativity by specifying "Events" in a coordinate-free, "invariant" formulation such as for S. It took Einstein 10 years, until 1915.

#21.7. CARTESIAN ORTHOGONAL COORDINATES OF LORENTZ EQUATIONS.

The Lorentz equations transformed from x to x′ and ct to ct′ in hyperbolic space can be modified to the preferable Cartesian orthogonal coordinates to using the substitution of ct = iT. The Lorentz equations (with the unprimed and primed variables interchanged, which change the sign of v) are:

$$x = [\,(x') - (vt')\,]\,[1 - (v/c)^2\,]^{-1/2}$$

Let   ct = iT and t = iT/c

Then $x = [\,(x') + (ivT'/c)\,]\,[1 - (v/c)^2\,]^{-1/2}$

$$ct = [(ct') + (vx'/c)\,]\,[1 - (v/c)^2\,]^{-1/2}$$

Also $iT = [(\,iT'\,) + (vx'/c)\,]\,[1 - (v/c)^2\,]^{-1/2}$

       $(-i)iT = (-i)\,[(\,iT'\,) + (vx'/c)\,]\,[1 - (v/c)^2\,]^{-1/2}$

Then   T = [(\,T' - (ivx'/c)\,]\,[1 - (v/c)^2\,]^{-1/2}

As explained below, parameters x, y, z and T represent the Cartesian universe (vector-spacetime) that we live in better that x, y, z and t. To do this, it is desired to put the equations for x and T in the form for conversion from one set of orthogonal Cartesian coordinates (unprimed) to another (primed) using the following relationships from analytical geometry, trigonometry, and linear algebra:

Coordinate Rotation by angle Ω:
    $x = x' \cos Ω - T' \sin Ω$
    $T = x' \sin Ω + T' \cos Ω$

Pythagorean Theorem:
    $x^2 + y^2 + z^2 + (T)^2 = x'^2 + y'^2 + z'^2 + (T')^2 = S^2$

Trigonometric Identities:
    $\sin^2 Ω + \cos^2 Ω = 1$
    $\sin ∅ = x/S$ $\sin ∅' = x'/S$
    $\sin θ = T/S$ $\sin θ' = T'/S$

Where Ω is the angle of rotation between the two coordinate systems, θ is angle between x and S, $\emptyset$ is the angle between T and S, θ' is angle between x' and S, $\emptyset'$ is the angle between T' and S, and so on.

To complete the conversions, the following substitutions are made:

$$\cos \Omega = [1 - (v/c)^2]^{-1/2}$$
$$\sin \Omega = (-i\,v/c) \cos \Omega = (-iv/c)\,[1 - (v/c)^2]^{-1/2}$$
$$\tan \Omega = \sin \Omega / \cos \Omega = -iv/c$$

The -i indicates the T coordinate axis is imaginary which is of no concern. In the linear algebra being used here, there are an infinite number of real axes and also an infinite number of imaginary axes possible all on equal footing.

,

$$x = [(x') + (ivT'/c)]\,[1 - (v/c)^2]^{-1/2}$$
$$= x' \cos \Omega - T' \sin \Omega$$

$$T = [(-(ivx'/c) + T']\,[1 - (v/c)^2]^{-1/2}$$

$$= x' \sin \Omega + T'\cos \Omega$$

*This is just what is wanted. The vector-spacetime x, y, z, and **T** we live at least as far as relative motion is concerned in is just a simple orthogonal coordinate system. The coordinate systems of observers moving at constant velocities with respect to each other find the values of the four vector-spacetime parameters x, y, z, and **T** vary for other observers based on a rotation of the other coordinate systems equal to an angle determined by their relative velocity v where tangent of the angle of rotation Ω is equal to v/c.*

*So whose coordinate system is rotating? Each observer says, "Not mine! Mine doesn't rotate, yours does." In fact, the coordinate systems don't rotate at all. The parameters x', y', z', and **T'** do, however, vary in magnitude for all the other observers based on their relative velocities.*

THE UNIVERSE, SPACE, AND BEYOND

#21.8. VECTOR-SPACETIME GRAPHS AND COORDINATE ROTATION.

It is interesting to graph the vector-spacetime (ct) and vector distance (x) and various aspects of special relativity: t is in units of seconds. Both ct and x are in units of the distance that light travels in one second (about 3 x 10^8 meters per second). It's preferable to graph ct as the vertical axis and x as the horizontal axis 90 degrees apart.

(1) For a beam of light moving at velocity c, relative to both a stationary observer and a constant velocity moving observer: When t=0, ct = 0, x = 0; when t = 1, ct = 1, x = 1; etc. Thus light travels the distance ct in in the x (horizontal) direction in each second and a distance ct in the t (vertical) direction in each second.

t	ct	x	
0	0	0	
1	1	1	
2	2	2	ETC.
-1	-1	-1	
-2	-2	-2	ETC

This plots as a line of 45 degrees on the graph. This represents how a beam of light looks to a stationary observer. It also represents the fastest that anything can travel since nothing can travel faster than the speed of light c. A negative t doesn't mean that time runs backwards, it just means that the beam of light is coming toward the observer's origin from the –x direction and then receding away from the origin in the positive x direction at the speed of light.

(2) Now superimpose on this graph how the coordinate system of an observer (x')
traveling half the distance x in each second looks to a stationary observer:

$$x = x' [(1- (v/c)^2)^{1/2} = 0.5 x$$

t	ct	x'	
0	0	0	
1	1	0.5	
2	2	1	ETC.
-1	-1	-0.5	
-2	-2	-1	ETC

This plots as a line through the origin at 27.5 degrees from the vertical ct axis.

Now visualize how the coordinates of the moving observer look to the stationary
observer. Assume the two observers (stationary and moving) start out with origins
aligned. As the moving ct' coordinate recedes, it will appear to the stationary observer
to rotate clockwise around the origin of the two coordinate systems compared to the
vertical stationary ct coordinate. Since the origins of the two coordinate systems (ct and
ct') are fixed together, for every second that goes by, the ct' coordinate is a horizontal
(x) distance ct further away from the origin as a result will appear to have rotated 27.5
degrees.

Exactly the same thing takes place for the x' coordinate except it will appear to rotate
counter-clockwise 27.5 degrees. Note that the direction of the two rotations is toward
the trace of an observer moving at the speed of light plotted in (1) above.

As the velocity of the moving observer increases, the rotation increases until the
velocity reaches the speed of light c and the rotation of the ct' and x' coordinates
reaches 45 degrees at which time the ct' and x' coordinates merge. At this point, the
velocity of the moving observer is measured at the speed of light by the stationary
observer while the moving observer measures stationary distances to be zero and the
stationary observer will find the moving clocks will have stopped.

#21.9. EXAMPLE OF COORDINATE TRANSFORMATION.

As an example, take two coordinate systems moving at a constant velocity $v = 0.6c$ relative to each other passing each other at exactly time $t = 0$ for clocks in both systems. A spacecraft launches at event S_1 at $ct_1 = ct'_1 = 0$ at $x_1 = x'_1 = 0$.
The spacecraft lands at event S_2 at $ct_2 = 30$ and $x_2 = 20$. Calculate unknown parameters S_1, S_2, x'_2, and ct'_2, T_1, T'_1, T_2, and T'_2.

$x_1 = 0$ $x'_1 = 0$

$ct_1 = 0$ $ct'_1 = 0$

$x_2 = 20$ $x_2 = ?$

$ct_2 = 30$ $ct'_2 = ?$

$ct = iT$

$T_1 = 0i$ $T_2 = 30i$

$(S_1)^2 = (x_1)^2 + (T_1)^2 = (x'_1)^2 + (T'_1)^2 = 0^2 + 0^2 = 0^2$ $S_1 = 0$

$(S_2)^2 = (x_2)^2 + (T_2)^2 = (20)^2 + (30)^2 = 1300$

Using the length formula in **section 21.6,**

$x'_2 - x'_1 = (x_2 - x_1) [1 - (v/c)^2]^{1/2} = [20 - 0] [1 - (0.6c/c)^2]^{1/2}$

$x'_2 = (20 - 0) (0.64)^{1/2} = (20) (0.8) = 16$

Thus, the moving observer would measure the distance traveled by the spaceship from start to the stationary landing point to be 16 (distance contraction) which less that the distance of 20 measured by the stationary observer.

The moving observer's clock reading t'_2 at the time of touchdown of the spaceship at t_2 is more difficult to calculate. The location of the spacecraft at event S_1 at $t'_1 = 0$ is exactly known as $x'_1 = 0$. The location of the spacecraft at event S_2 at $t'_2 = 0$ is also exactly known as $x'_2 = 16$. The elapsed time **on the moving observer's clock at t_2** is more difficult to derive. One way is to calculate the time on the moving observer's clock t' up to t_2 and then include the time it takes light to catch up to the spacecraft after t_2. Fortunately, there is an easier way.

The **moving observer's clock reading ct'$_2$ at t_2** can be found from the equation:
$$(S_2)^2 = (S'_2)^2 = 1300 = (x'_2)^2 + (T'_2)^2$$

$$(T'_2)^2 = 1300 - (x'_2)^2 = 1300 - (16)^2 = 1044$$
$$(T'_2) = 32.31$$

Note that this is not the result that would be obtained for t' by using the Lorentz equation for a moving observer t'. (t') would calculate the time t **on a stationary observer's clock**. T' is the time that was on **the moving observer's clock t'** when the spaceship touched down ($S_2 = S'_2$) at the stationary observer's x_2 at time t_2. S_1 and S_2 are "invariants" that remain the same for all observers regardless of their relative motions.

#21.10. A QUESTION OF SIMULTANEITY: WHEN WAS THE SHOT FIRED IN NEW YORK CITY?

Assume you live in Los Angeles (LA) and want to investigate when a shot was fired in New York City (NYC). Further, assume you are a space traveler traveling at 0.6c and wish to know where and when the crime took place.

To find the distance to NYC, put a mirror at NYC and bounce a light beam from LA to NYC and back to LA. Time the light beam and find the light beam traveled distance of 6000 miles to NYC and back. So the distance from LA to NYC is 3000 miles for a stationary observer.

As the space traveler, fly directly over LA synchronizing your clocks to zero and continue directly to NYC. You find the distance to NYC as determined from the Lorentz transform is:

$$L' = L \ [1 - (v/c)^2]^{1/2} = 3000 \ [1 - (0.6)^2]^{-1/2} = 3000 \ (0.8) = 2400 \text{ miles}$$

If you were in **NYC** when the shot was fired, there would be no difference in your appraisal when the shot was fired. If you were in **LA,** you would find it happened later by the time it takes light to travel to **LA**. If two shots were fired, you would all agree on the interval (Proper Time T) between the shots as proper time is an invariant.

#21.11. IS "T" or "t" A FOURTH DIMENSION?

It's easy to jump to the conclusion that T and t are some kind of a "fourth" dimension. It's also easy to say the space rotates in Cartesian (x, y, z, T), polar (r, Ω, θ ,), hyperbolic (sinh, cosh), or whatever coordinates depending on the mathematical formulation (t or T) used to describe observers traveling at different constant velocities approaching the speed of light.

Yes, yes, yes, of course time is a fourth dimension! We live in a spacetime universe and spacetime of four dimensions. If you want to meet someone, you must specify the time (t) coordinate as well as the address coordinates (x and y) and sometimes the building floor coordinate (z).

We all have always lived in a four-dimension world. Light whizzes by us all day long at the speed of light with no obvious rotation of vector-spacetime. In total darkness, as far as we can tell, there is no change in the rotation of vector-spacetime.

Every time you walk out the door and return, you have aged less than those you left behind. Not much, maybe a few micro-micro-nanoseconds, depending on how long you were gone and how far and fast you traveled. When relative velocities begin to approach an appreciable percentage of the speed of light though, then i**ct** or **T** must be specified, in addition to x, y, and z. The specific values of all four coordinates in other frames will have to be translated to the coordinates of an observer traveling at a different constant velocity using the Lorentz coordinate transformations. The Lorentz equations show that a variation in the vector-spacetime (ict and **T)** coordinate in one frame may change both the vector-spacetime and vector distance x, y, and z coordinates in another; and likewise, the distance coordinates in one frame may change both the time and distance coordinates in another.

If an object is coming toward you (or moving away from you) at a constant velocity approaching light velocity, then, the object will appear shorter to you (and its clock would run slower than yours) although it would whiz by so fast you wouldn't even see anything; if a large spaceship were to hit the Earth at that velocity, it might destroy the Earth (as well as the spaceship)!

Some physicists speak of the past, present, and future existing in the Universe all at once and maybe forever. That's just stretching it. Your present exists for you wherever you are at the moment it happens. Your present doesn't exist until then anywhere else; and even then, it doesn't exist anywhere else until the "news" has "time" to reach there (anywhere else) perhaps long after it happened (according to your clock) where you are. The quickest it can reach anywhere else is limited by the speed of light (the velocity that modulated electromagnetic waves can travel).

Sure enough, the news of your present happening can continue outward on its journey literally forever. So does your past, too, continue to exist forever. How long it takes for an "Event" of your past, present, or future to "catch-up" to anywhere else in the Universe is a matter for the Lorentz transformation equations to reconcile. Rest assured though, that no matter how fast other inertial frames travel with respect to your inertial frame, you will be quite comfortable watching them whiz by even at near the speed of light as long as they don't crash into your inertial frame, although if you blink, you may miss seeing them, or even if you don't blink, they may whiz by so fast you can't see them. Your personal clock will just keep ticking normally. Those in the other inertial frames will be just as comfortable as you are in yours and their clocks will just keep ticking normally too although much slower than yours.

And, they all will say that your distances are contracted and your clock is running slow. "That's just the way it is."

As they are whizzing by, however, their length will be contracted and distorted accordingly (length contraction) by the amount specified by the Lorentz equations.

#21.12.   A TIME MACHINE (THE TWIN PARADOX).

If you, an astronaut, travel away and return at very high velocities, close to the velocity of light, you would find that the reference clock (and your twin) you left at home have gained perhaps dozens or even hundreds of years compared to you and your personal clock that you have with you. Your personal clock may have gained only a few years. This is called the twin paradox.

As an astronaut, if you travel near the Earth at 1 mile per second, that would be 3600 miles per hour. That's still pretty fast! At that speed, it will take 2,232,000 years to reach a star that is 12 light-years away. You know that you have to bump it up a notch say to 0.6c (0.6 of the speed of light). That's 111,600 miles per second.

If you would travel to a distant star 12 light-years away at 0.6c and return, your twin and the folks left at home would have aged 40 years.   You, on the other hand, would have aged only 32 years. How is this possible you might inquire? It seems that with respect to you, your stationary twin and the people left behind at home were moving away from you at 0.6c and then toward you at 0.6c. So their clocks should run slower than yours and they should be younger than you (or maybe the same age)? Who is right? Is your twin left at home younger or are you the traveler?

You, the traveling twin is younger. It's because of time dilation, length contraction, and acceleration, and deceleration. All will give the same result, but it is interesting to look at all of these in the explanations in the next four **section**s.

#21.12.1. THE TRAVELING TWIN IS YOUNGER DUE TO TIME DILATION.

First, consider time dilation. The astronaut twin traveled at (an average) of 0.6c about 111,600 miles per second (179,564 kilometers per second). Using the Lorentz formula, the astronaut twin's clock will run slower than the stationary clock of the twin at home.

EXAMPLE OF TIME DILATION: The twin left on Earth will measure the moving astronaut twin's clock running slower compared to stationary clocks: as follows

$$T_{MovingTwin} = T_{StationaryEarth} \times [1 - (u/c)^2]^{-1/2} = T_{StationaryEarth} \times [1 - (0.6c/c)^2]^{-1/2} =$$

$$T_{MovingTwin} = T_{StationaryEarth} \times [1 - (0.6)^2]^{-1/2} = T_{StationaryEarth} \times [1 - (0.36)]^{-1/2}$$

$$T_{MovingTwin} = T_{StationaryEarth} [0.64]^{-1/2} = T_{StationaryEarth} (1/0.8) = 1.25 \times T_{StationaryEarth}$$

For every 1.25 hours the stationary Earth clocks advance, the moving twin's clock advances 1.0 hour.

$$T_{StationaryEarth}] = [0.8] \times T_{MovingTwin}$$

Similarly for every 0.8 hours the moving twin's clock advances, the stationary earth clocks advance 1.0 hour.

The stationary clock of the stationary twin indicated that the round trip of the astronaut twin to the distant star took 40 years. This is 1.25 times longer than the astronaut twin's clock indicates. The astronaut twin's clock indicates that the round trip took 32 years, 0.8 times less.

#21.12.2. TRAVELING TWIN IS YOUNGER DUE TO DISTANCE CONTRACTION.

It's even more interesting to explore the astronaut twin's travel from the standpoint of distance contraction. The twin at home being stationary will measure the distance to the "stationary" distant star to be 20 light-years. Even though the twin at home and the distant star are a long way away from each other, they are stationary with respect to each other. It would take 20 years for light to travel from the stay-at-home twin to the distant star and 20 years for light to travel from the distant star back to the stay-at-home twin.   A pulse of light sent to and reflected from the distant star would take 40 years to make every round trip. If the distant star and stay-at-home twin's clock were sending out light pulses every hour, each successive hourly pulse would arrive at its destination exactly one hour after the previous one measured by the stationary clocks on the distant star and on the Earth.

On the other hand, the astronaut will notice on the journey to the distant star that each successive pulse of light from the star arrives in less than an hour as measured by his personal clock on board the Spaceship. The astronaut is traveling toward the distant star and is closer to the star for each successive light pulse. (The astronaut's clock is running slower than the clock at the distant star.) The faster the Spaceship travels with respect to the distant star, the faster the light pulses arrive after each other and the slower the astronaut's clock is measured to run with respect to the clock at the star.   In fact, upon arriving at the distant star, the astronaut would have counted 20 years of hourly pulses from the distant star but the astronaut's clock would show much less time for the trip as it is running slower than the stationary clock on the distant star.

As the astronaut voyages toward the star, the distance the hourly pulses must travel to the astronaut is less. The first pulse traveled 20 light-years. The next one travels a lesser distance because the Spaceship is closer to the distant star. The last one travels no distance at all as the traveler has arrived at the star. In all, it took 16 years for the astronaut to arrive at the distant star.

A similar situation takes place as the astronaut turns around and heads home at an (average) velocity of 0.6c. The hourly light pulses from home arrive less than an hour apart based on the astronaut's clock. Upon arriving home, the astronaut would have counted 20 years of hourly pulses from home, but the astronaut's clock would show it was running much slower and indicated another 16 years.

The initial light pulse from the distant star takes 20 years to arrive at the Spaceship just as the astronaut leaves the Earth. Each successive light pulse is received in less than an hour by the astronaut's clock as the Spaceship travels toward the star and light travels less distance to the Spaceship. The last light pulse from the distant star arrives at the Spaceship just as it docks on at the distant star and is received instantly.

The astronaut's travel time back to the Earth is the same. The initial light pulse from the Earth takes 20 years to arrive just as the astronaut leaves the distant star to return to the earth. Each successive light pulse from earth is received in less and less of an hour by the astronaut. The last light pulse from the Earth arrives at the Spaceship just as it docks on Earth and is received instantly.

EXAMPLE OF DISTANCE CONTRACTION:

$$L_{MovingAstonaut} = L_{StationaryEarth} \times [1 - (u/c)^2]^{1/2} = L_{StationaryEarth} \times [1 - (0.6c/c)^2]^{1/2}$$

$$= L_{StationaryEarth} \times [1 - (0.36)]^{1/2} = L_{StationaryEarth} \times [0.64]^{1/2}$$

$$L_{MovingAstonaut} = L_{StationaryEarth} \times [0.8] = 20 \text{ light-years} \times 0.8 = 16 \text{ light-years}$$

The astronaut traveled a distance of 16 light-years to reach the star and 16 light-years to return, 32 light-years in all. The stationary twin and others on earth, and those stationary at the distant star, will find by their measurements that the astronaut twin traveled a distance of 40 light-years and they (the twin on Earth, people on Earth, and beings at the distant star) have aged 40 years.

The twin on the Earth would measure the astronaut twin's clock to only have advanced 0.8 x 40 years = 32 years when the astronaut twin returned and set foot on the Earth again.

#21.12.3. TRAVELING TWIN IS YOUNGER DUE TO ACCELERATION "LOCKING- IN" NEW (SLOWER) CLOCK RATES.

Both the astronaut twin and the Earth twin think the other's clock runs slow.
Who is correct?

The Earth-bound twin is correct! The Earth-bound Twin has aged 40 years while the astronaut twin has aged only 32 years.   The astronaut twin cannot claim that the clocks on earth have run slower than his because he was not at a constant velocity during periods of acceleration and deceleration at the trip's beginning, midpoint, and end. This lack of uniformity removes the constant velocity symmetry between clocks of the astronaut twin and the Earth twin. The astronaut twin's calculations do not take into account that his clock and constant velocity frame (location) have been subjected to acceleration. Careful inclusion of these effects resolves this conflict and is part of Einstein's General Theory of Relativity. See **section 22.**

To be entirely accurate, the effects of acceleration are really what are essential for your time machine. When the effects of acceleration are included, as explained in the next section on General Relativity, a more accurate picture emerges:

When relative velocity is a constant, the Earth and Astronaut Twins both correctly measure each other's clocks to run slower as determined by the Lorentz Transform and both twins are correct in this regard.

However, when the effects of acceleration are taken into account by the equations of General Relativity, the sustained bursts of Astronaut Twin's acceleration will extensively slow the Astronauts clocks by much more than determined by the Lorentz Transforms. Both twins will agree if this is taken into consideration. These bursts of acceleration will be enough to further slow the astronaut twin's clocks and will more than make up for the Earth twin's clocks running slower than the astronaut's clocks during periods of constant velocity.

#21.12.4. ACCELERATION AND DECELERATION OF TRAVELER'S CLOCK TAKE TRAVELER'S CLOCK (AND TRAVELER WITH IT) TO NEW (SLOWER) CLOCK RATE.

The effect of acceleration on a clock is to make it run slower. Deceleration will then make a clock run faster. It takes force (such as some kind of propellant or gravitational acceleration) to accelerate a previously constant-velocity clock and make it run slower. The acceleration of the moving twin was enough to greatly slow their moving clock so it ran much, much slower that the stationary twin's clock. This slowing greatly slowed its rate compared to the stationary clock. The effect was that the moving twin's clock now ran much, much slower than the stationary twin's clock After the periods of acceleration and deceleration, the moving twin's clock would return to its normal rate and both twins would find the other's clock ran slower.

So, if it were possible, how fast can you travel and how slow can you make your personal clock run compared to your reference clock at home? The answer is fast enough so that your clock slows down and (almost) stops entirely with respect to the reference clock at home. The velocity at which this happens is the velocity of light which is 186,000 miles (299,000 kilometers) per second. For you travel this fast would take an infinite amount of energy to get you up to speed-of-light velocity. At speed-of-light velocity, you would be converted to pure energy. (Ouch!)

The onboard clocks carried by light photons, other electromagnetic-energy photons, gravity's gravitons, de Broglie waves, and subatomic particles that travel at or near the velocity of light do find that all the other clocks in the Universe and Spacetime beyond are running much slower than their clocks and have almost stopped, or have stopped. With all other clocks stopped, they have plenty of time on their onboard clocks to travel anywhere instantly in this Universe and Space beyond (as measured by the stopped clocks). There is more about this in the following **sections**.

#21.13. EQUATIONS OF PHYSICS MUST BE MODIFIED TO INCORPORATE SPECIAL RELATIVITY EFFECTS.

The following material provides a summary of how Special Relativity changes the equations for velocity, momentum, and energy.

MOMENTUM (p)

Momentum (p) in the non-relativistic formulation, $p = mv$, where m is mass and v is the velocity of the mass, is an important quantity because momentum remains constant (is conserved) in a collision (interaction) between two objects. When collisions occur at velocities approaching the speed of light, momentum is not conserved unless the equation $p = mv$ is modified. The **bold** print indicates the quantity is a vector. The correct relativistic formulation of the momentum equation for which momentum is conserved is:

$$p = m_0 v [1 - v^2/c^2]^{-1/2} = m_r v \text{ where } m_0 \text{ is defined as rest mass.}$$

Note that $[1 - (v/c)^2]^{-1/2}$ is simply the parameter from the Lorentz transform and varies from 1 to infinity as v varies from zero to infinity.

Relativistic mass m_r is defined as $\quad m_r = m_0 [1 - v^2/c^2]^{-1/2}$

The above equation for relativistic mass m shows that as velocity (v) increases to near the speed of light (c), relativistic mass m_r approaches infinity. That's why nothing can exceed the speed of light.

FORCE (F)

In non- relativistic formulation for the force F on a mass (m) due to acceleration a:

$$F = ma = mdv/dt = dp/dt$$

Force here is the rate of change of momentum (dp) with respect to time (dt) and applies to both non-relativistic and relativistic motion.

Acceleration (a) is the rate of change of velocity (dv) with respect to time (dt)
$$a = dv/dt$$

TOTAL ENERGY (E)

$$E = E_0 + K$$

Where E_0 is the rest energy; K is the kinetic energy.

E is defined as $E = m c^2$ (Einstein's equation). **For energy to be conserved, relativistic mass m_r must be used. Also, any other energy taken on by the mass (heat, pressure, potential, etc.) must be accounted for as well.**

Or, $E = m_r c^2 = m_0 c^2 [1 - v^2/c^2]^{-1/2}$

This equation can be solved using the binomial theorem (**section 12.2.1**):

The first four terms in the series are:

$$(a + u)^n = a^n + na^{n-1} u + n(n-1) (a^{n-2} u^2) /1 \times 2$$

$$+ n(n-1) (n-2) (a^{n-3} u^3) /1 \times 2 \times 3 + ...$$

Thus, for the equation for relativistic mass: $m_r = m_0 [1 - (v/c)^2]^{-1/2}$

$a = 1,$ $u = - (v/c)^2,$ and $n = - 1/2$

The first three terms for relativistic mass m are:

$$M_r = m_0 [1 - (v/c)^2]^{-1/2} = (m_0) [1^{-1/2} + (-1/2)(1^{-1})(-v/c)^2$$

$$+ (-1/2)(-1/2 - 1) (1^{-1/2 - 2})(-v/c)^4 /1 \times 2 + ...]$$

Thus, $m_r = m_0 [1 + (1/2)(v/c)^2 + (3/8)(v/c)^4 + ...]$

Substituting this into Einstein's equations for energy E:

$$E = m_r c^2 = m_0 c^2 [1 + (1/2)(v/c)^2 + (3/8)(v/c)^4 + ...]$$

21.13. EQUATIONS OF PHYSICS MUST BE MODIFIED TO INCORPORATE SPECIAL RELATIVITY EFFECTS (CONTINUED).

This result is interesting for several reasons: The first term ($m_0 c^2$) is simply Einstein's term for the equivalence of mass and energy – a little bit of rest mass m_0 can be converted to a lot of energy. It is called the rest mass or rest energy of matter. The second term ($1/2\, m_0 v^2$) is the non-relativistic kinetic energy of velocity of mass, its non-relativistic kinetic energy (K). The third [$(3/8)(m_0\, v^4 c^2) + ...$] and subsequent terms are additional energy due to the relativistic effects described in this **section**.

The total energy (E) is the sum of the rest energy (E_0) plus the relativistic energy and non-relativistic kinetic energy (K). An object in motion also has additional gravity due the equivalent mass of this kinetic energy. **Any other energy taken on by the mass (heat, pressure, potential, etc.) must be accounted for as well.**

 This is a statement that mass and energy are two forms of the same thing. This is why the equivalent mass of pure energy such as of a photon is attracted by other masses. Thus, light from distant galaxies and stars is bent by the Sun's gravity as it travels by on its way to the Earth. Experiments show however, the amount of bending predicted just on this basis alone is not correct and is only about half of the actual value. This is due to other effects of acceleration of various kinds and other types of energy. Einstein's General Relativity explains the actual value of the bending of light and other phenomena of gravity and acceleration as accurately predicted by Einstein's General Relativity equations. (See **section22**.)

KINETIC ENERGY (K)

If an object is initially at rest, the work done on the object as it accelerates is equal to its kinetic energy (K).

$$K = m_0 c^2 (1-v^2/c^2)^{-1/2}$$

For an object **at rest**, $K_{rest} = 0$

For **non-relativistic** kinetic energy $K_n = \frac{1}{2} mv^2$ of an object moving at 0.9c, $K_n = 0.45$ mc^2

For **relativistic** kinetic energy $K = mc^2 (1-v^2/c^2)^{-1/2}$ of an object moving at 0.9c,
$$K_r = 1.29 \ mc^2$$
Using non-relativistic equation would result in an error of over 300 percent!

APPLICATION OF CONSERVATION OF RELATIVISTIC ENERGY AND MOMENTUM

A particle of mass M decays into two identical particles m and m: Energy is conserved, use the formula for energy:

$$E = mc^2 / (1 - v^2/c^2)^{-1/2} \quad M c^2 / (1 - 0^2/c^2)^{-1/2} = 2mc^2 / (1 - v^2/c^2)^{-1/2}$$

$$M = 2m/ (1 - v^2/c^2)^{-1/2} \quad v = c [1- (2m/M)^2]^{1/2}$$

For M to decay into two particles, M must be equal or greater than 2m. If M is greater than 2m, any extra energy will be dissipated by giving extra kinetic energy in the form of additional velocity to the two particles m. Note that momentum is conserved this interaction. If M is initially at rest, m and m go off in opposite directions with equal velocity and momentum. If M is not at rest, then its relativistic momentum and energy will be shared equally by m and m.

#22. GENERAL RELATIVITY.
#22.1. EINSTEIN'S SPECIAL RELATIVITY AND GENERAL RELATIVITY ARE VERY DIFFERENT.

The effects of Special Relativity and the constant value of the velocity of light are not noticed and are of no concern to most people on the Earth. Those early scientists who learned that the velocity of light was a constant regardless of the relative velocity of the source or an observer were surprised and perhaps startled. The ramifications were not understood even after Maxwell's equations predicted in 1865 that the velocity of electromagnetic waves (photons) was a constant (the velocity of light) for all frames of reference of sources and observers. Various explanations were put forward including Lorentz transforms and luciferous (solid) ether. It took an Einstein to clarify matters in 1905 with his Special Relativity.

Compared to the complexities of General Relativity, the mathematics of Special Relativity are straight forward and uncomplicated. With a simple substitution $(-ct)^2 = (T)^2$ where (T = spacetime) in the Pythagorean theorem $(x^2 + y^2 + z^2 + ... = h^2)$ for the length of the hypotenuse of a (multi-dimensional) right triangle, the familiar high school trigonometry and linear algebra of right triangles and Cartesian coordinates can be extended and used to solve many Special Relativity problems related to clocks running slower and distances contracting (spacetime warping) as relative velocity increases (**section 21.4**).

$$x^2 + y^2 + z^2 + (T)^2 = S^2 = x'^2 + y'^2 + z'^2 + (T')^2$$

Also, the equations of physics are straight forward to modify for Special Relativity, usually by including the Lorentz transform $[1 - v^2/c^2]^{-1/2}$ such as for relativistic mass (**section 21.13**):

Relativistic mass m_r is defined as $m_r = m_0 [1 - v^2/c^2]^{-1/2}$

The above equation for relativistic mass m shows that as velocity (v) increases to near the speed of light (c), relativistic mass m_r approaches infinity. It would take infinite energy to accelerate an object to the velocity of light. That's why "nothing can obtain a velocity that exceeds the velocity of light."

Various (perhaps dubious) explanations have been theorized for seeming exceptions exceeding the velocity of light during our early inflating Universe and our aging expanding Universe. The explanations described in this book are:

- Expansion is not in an inertial frame.
- The "Higgs fields were frozen during inflation of our early Universe and are frozen beyond our Universe.

However, Special Relativity **(section 21)**, as powerful as it is, applies only to constant velocity (unaccelerated) motion. To be complete, accelerated motion is investigated in Einstein's General Relativity and described this section including motion of objects influenced by gravity or any other kind of acceleration. Since energy is equivalent to mass, both mass and energy must be considered as contributing to the geodesic of motion of an accelerated object.

The basic qualitative effects of gravity have been known in intimate detail to almost everyone who ever lived on Earth. The fact that heavy and light objects dropped side-by-side move at the same velocity and acceleration has been studied and known for thousands of years. Most everyone with a rudimentary education knows that gravity has caused the Moon to orbit the Earth and the Earth to orbit the Sun for billions of years and will cause them to keep orbiting (we all hope) for billions of years more.

The mathematics of Sir Isaac Newton in 1679 gave very (almost) accurate equations that describe the effects of gravity on objects in **freefall**, in **orbit**, or **with escape velocity**. Newton's laws and equations **suffice for most applications even today**. The equations of Einstein's General Relativity are necessary only in very extremes of mass, energy, and acceleration.

The reader should note carefully **the tiny errors** in Newton's equations compared to Einstein's equations in the following two examples: Newton's equations predict a perturbation of the orbit of Mercury of 531 **seconds of arc PER CENTURY** (!), but Einstein's equations agree with experiment predicting 574 [Lieber p. 275]. Similarly, Newton's equations predict the bending of light passing near the Sun about to be about 0.87 **seconds** (!), but Einstein's equations agree with experiment predicting 1.75 [Lieber p. 287].

#22.2. <u>QUANTITATIVE MATHEMATICAL TREATMENT OF GENERAL RELATIVITY.</u>

Unlike the mathematics of Special Relativity, the mathematics of General Relativity is complex and to some if not most "incomprehensible" [Collier, Front Cover].

A comprehensive quantitative mathematical treatment of General Relativity in this book would add hundreds of pages that duplicate information that is readily available elsewhere and might not be of interest to most readers.

Three books which provide the necessary quantitative information are available at very reasonable cost ($10 to $20). These books are recommended and listed in the **Bibliography** as follows:

Lieber, Lillian R, *The Einstein Theory of Relativity,* Paul Dry Books (2008) - 350 pages. This is a very good book to start with and then progress to the others in the order listed.

Collier, Peter, *A Most Incomprehensive Thing: Notes Toward a Very Gentle Introduction to the Mathematics of Relativity* (2012), 364 pages. (

[1] McMahon, David, *Relativity Demystified*, McGraw-Hill (2006) - 345 pages.

Accordingly, this section concentrates on a discussion of qualitative aspects of General Relativity and includes mathematics as necessary for clarity.

#22.3.1. ENERGY IN SPACETIME IS EQUIVALENT TO MASS.

Mass is equivalent to energy according to Einstein's equation: $E = mc^2$. A little mass makes a lot of energy. A lot of energy makes a little bit of mass – this LITTLE BIT OF MASS must be considered to accurately calculate geodesic paths of motion.

Rest mass is the mass of a particle at rest. This is basic energy of particles, hadrons and atoms.

Adding energy (heat, pressure, mechanical, potential, kinetic, relativistic, stress, etc) adds mass ($m = E/c^2$)

The energy and mass of particles, hadrons, and atoms is conserved and can be released in particle interactions through decay, fission, and fusion.

 Rest mass m_o increases with velocity according to the Special Relativity equation for relativistic mass $m_r = m_o [1 - (v/c)^2]^{-1/2}$ (**sections 5.3 and 5.4**).

The Cartesian spacetime of Special Relativity (**section 21**) is found only where objects are not subject to acceleration. In Cartesian spacetime, an unaccelerated object moves in a geodesic of a straight line at a constant velocity.

#22.3.2. SPACETIME CURVE IS GEODESIC.

Mass and energy cause spacetime to curve (warp) into a geodesic (the shortest distance curve between two points in spacetime). These warps and curves create an effect on the motion on objects called Gravity. A geodesic so created is the path any **freefalling** object there will follow in spacetime.

Contrary to Newton's equation (**section 1.2.3**), Gravity is not a Force. Gravity doesn't exist as an entity. Gravity is an effect of mass and energy altering the geometry of spacetime (warping spacetime) into geodesic paths of the shortest geodesic distance between two points in spacetime. **Objects in a geodesic are FREEFALLING and follow the geodesic in which they are moving perhaps under the impetus of gravitons. Their motion in the geodesic gives the appearance and effect of what is called Gravity.**

If a spaceship is accelerated, objects (that are not fastened down) would drift freefall to the rear until they encounter the rear end of the spaceship. This is similar to jumping off a bridge and drifting freefall until you encounter the ground (Ouch!).

A freefalling observer far from any source of Gravity cannot tell any difference in freefalling and being accelerated under the influence of a massive object such as the Earth or the Sun.

A freefalling observer cannot detect being accelerated even in the presence of a massive object. (Under the influence of extremely massive objects such as black holes, an observer freefalling into a black hole might experience a huge difference in acceleration between head and feet and could be torn apart.)

For two observers **freefalling** (in a vacuum so air resistance is not a factor) **in the presence of a massive object** with one observer overtaking the other, both observers would claim that the other is moving while they are stationary. This is because both observers are subject to the same acceleration, for example, 9.8 meters/sec^2 near the surface of the Earth. If either observer though moving at different velocity would drop an object, it would hover and continue to fall at the same rate as the observer who dropped it.

For two observers free falling NOT in the presence of a massive object with one observer overtaking the other, both observers would claim that the other is moving while they are stationary. This is because both observers are subject to the same acceleration: ZERO. This is the constant relative velocity situation covered in Special Relativity (**section 21**).

All objects freefall at the same rate regardless of the magnitude of acceleration they are subjected to. All objects regardless of their composition (cannon ball, base ball, ping pong ball, towel, etc) freefall (in a vacuum) at the same rate regardless of the magnitude of acceleration they are subjected to.

In general, the effect of acceleration due to mass and energy cannot be distinguished from any other type of acceleration – elevator, rocket engine, and the like.

An observer standing on a stationary platform on the surface of the Earth or standing on an upward accelerating platform in empty space of the same acceleration as at the surface of the Earth (9.8 meters/sec^2) could not distinguish between the effects of the platforms on the observer or on objects which fell downward to the platform when released by the observer.

#22.4. ACCELERATION BY GRAVITY AND OTHER MEANS.

Free-falling masses (such as a baseball or cannon ball, or feather) free falling side-by-side at the same location acted on only by a gravitational field, will freefall (accelerate), side-by-side at exactly the same rate and have the same accelerated velocity regardless of the value of their masses. Any two objects dropped side-by-side on the Earth will hit the ground at exactly the same time discounting the effect of air by dropping the objects in a vacuum.

Gravitational-mass, inertial-mass, and energy-mass have exactly the same characteristics in a gravitational field or under other acceleration. Acceleration of an elevator going down counteracts the effect of gravity; going up it increases gravity. It is impossible to distinguish any difference in effect of acceleration due to gravity or some other form of acceleration.

Acceleration by any means has the same effect on the objects being accelerated as acceleration by the same amount acceleration due to mass and energy.

#22.5. CURVING (WARPING) OF SPACE TIME BY MASS AND ENERGY PLACES OTHERWISE UNACCELERATED OBJECTS IN GEODESIC ORBITS WHERE THEY FREEFALL.

An astronaut in a freefall space ship in orbit around the earth is essentially weightless and cannot detect the Earth's gravitational field. The astronaut, objects in the spaceship, and the spaceship itself are in freefall orbit due to the **GEODESIC ORBIT (spacetime warp) CREATED BY THE MASS AND ENERGY OF THE EARTH**. The spaceship is traveling in an orbit of warped (curved) spacetime and is following a geodesic (shortest spacetime distance) path caused by the warping (curving) of spacetime by Earth's mass and energy. Depending on the spaceship's velocity, it could (1) continue "freefall" in orbit around the Earth indefinitely just like the Moon, (2) eventually freefall and crash into the Earth if its initial velocity is small, or (3) if its initial velocity is large enough, freefall escape (**section 2.5.3**) into outer space and freefall forever free of theEarth's gravity unless "captured" by a mass-energy geodesic of some other large object.

A spaceship in permanent orbit must activate its engine to accelerate and increase velocity to escape from the clutches of the Earth's gravitational acceleration (**section 2.5.3**) or reduce velocity and change its orbit so as to return to Earth and land.

General Relativity provides a means to quantitatively express and calculate the motion of objects subject to effects of mass-energy, in our Universe from planets, stars, galaxies, black holes, and other celestial objects. Einstein's equations of General Relatvity calculate motion to accuracies beyond Newtonian mechanics (**section 1.2.3**).

If there were a tunnel to the center of the massive object, the massive object's effect would become less and less as you fell into the tunnel below the surface and would decrease to zero as you arrived at the center of the massive object- there would be an equal amount of mass all around you cancelling out the entire effect of mass-energy. (Nevertheless, the massive object would still be trying to pull you apart at its center, adding negative energy to you, and decreasing your weight.) If the tunnel came out the other side of the massive object, so would you. Then spacetime would pull you back in and you would come out the other end, and so forth - you would oscillate in and out of the massive object although, for a very massive object, the difference in acceleration at your head and feet may be enough to tear you apart).

#22.6. EFFECT OF GRAVITY AND ACCELERATION ON CLOCKS (TIME) AND DISTANCE.

Clocks run slower as the strength of the acceleration from mass and energy or other acceleration increases. Clocks on Earth run slower than those on the moon. Clocks on Jupiter run slower than clocks on Earth. Clocks in an infinite gravitational field stop.

Light from a source in a higher gravitational field is shifted toward the red. Light in lesser gravitational field is shifted toward the blue. Light from a receding source is shifted toward the red. Light from an approaching source is shifted toward the blue.

A clock near a massive object (m) such as the sun runs slow compared to clocks further away a distance r from the center of the sun. For example, for an object such as the sun of mass (m), clocks a distance r away from the run slower by:

$$(1 - 2mG/rc^2)^{1/2}$$

Where G is a gravitational constant (**section 1.2.3)** and c is the speed of light.

This equation was derived from Einstein's equations in 1916 by Karl Schwarzschild. See also [Stannard p. 82]. As the distance away from the massive object increases, clocks speed up and eventually will run normally (as r approaches infinity). Light is also slowed down and red shifted by the massive object.

Distances are also shortened by massive objects by the same factor above. From the perspective of a distant observer, the closer a spaceship gets to the massive object the shorter the spaceship becomes.

Light is also slowed down as it passes a massive object. Light is bent more as it slows down when it passes very near our sun much more than than would otherwise happen just from the equivalent mass of the sun and light photons.

As the light slows down, it has more time to bend toward the sun. This is not a violation of the universal velocity c of light. It is just like light slows down and is bent by water or a prism. There is not an inertial platform of constant velocity in these instances.

If you were falling toward a massive object, your clock would run slower and slower and distances would get shorter and shorter by the same factor:

$$(1 - 2mG/rc^2)^{1/2}$$

As you continue falling, at a distance $r = 2mG/c^2$, the above expression would reduce to zero and your clock would stop. This value of r is called the Schwarzschild radius. It specifies a spherical surface called the "event horizon." If all the mass m were squeezed inside a sphere of radius r, light could not escape and a black hole (**section 6.3**) would have been created. Planets and stars cannot shrink by themselves to black holes as their internal atoms exert counter pressure to prevent it. To make the Earth a black hole, it would have to be shrunk to the size of a small marble. However, if enough mass is collected, gravitational force becomes large enough to collapse the atoms and its constituent particles together to create a quasar and a black hole.

#22.7. GRAVITY AND ACCELERATION CREATE *GEODESICS* THAT PLACE OBJECTS IN FREE-FALL GEODESIC ORBITS.

Assume two objects in space: one heavy (massive) such as a spacecraft, and one light (very little mass) such as an astronaut untethered beside an unpowered spacecraft. Both float side-by-side freefalling untethered around the Earth. Both are following the same space geodesic. Both are free falling in a spacetime warp or geodesic. It would make no difference if they instead were falling toward the Earth. They still would do so untethered and fall side by side exactly together to the Earth. (**See section 13.7.**) Depending on the velocity of the spacecraft, there are three possibilities (with the astronaut moving along with the spacecraft):

1. The spacecraft has "escape velocity" and is in a geodesic in which it will continue to move away from the Earth, break free of the "clutches" of the Earth's gravity, and never return to the Earth.

2. The spacecraft does not have enough velocity to put it in orbit and it is in a geodesic that will eventually accelerate it due to the Earth's gravity and crash it into the Earth unless saved by its engines.

3. The spaceship has a velocity sufficient that its geodesic will keep it in orbit around the Earth. The force of gravity is not necessary. This spaceship will continue to orbit the Earth perhaps forever (unless it is ultimately slowed by friction of stray "space dust and then slow enough to crash into the earth). If its orbit is circular, it will stay a constant distance from the Earth. If its orbit is eccentric, it will accelerate as it nears earth and then decelerate as it move away from the Earth – never crashing into the Earth, but never escaping the earth's gravity

Case 3 above is stunning!!! It is at the heart of General Relativity. A spaceship and the Earth's moon are going to orbit the Earth perhaps for billions of years with no added energy or fuel! How can that be? How can the Moon orbit the Earth for billions of years? Where is the energy to keep it in orbit coming from?

The answer is that they are in geodesic orbits and once in a geodesic orbit, no energy is necessary to keep an object in this orbit. The geodesic orbit is the natural orbit that Space has warped to create FOR A FREEFALLING OBJECT. The geodesic orbits are the natural paths of motion of objects in spacetime due to gravity characteristic of mass and energy.

Objects follow scalar geodesic paths in response to the warping of space due to mass and energy. If moving at a constant velocity, an object will continue on at the same velocity unless disturbed by the **GEODESIC OF MASS AND ENERGY** causing it to accelerate (or decelerate).

Mass and acceleration warp (curve) spacetime. Spacetime geodesics are the warps (curves) that objects follow in a Higgs scalar gravitational field (**section 13.7.7**)

All objects regardless of mass have the same freefall acceleration in a geodesic and follow a spacetime warp geodesic curve side by side. Think of yourself and everything on Earth or going around the Earth as being in a freefall geodesic orbit of the Earth. Carefully jump up off a short step – or better yet, watch someone jump out of an airplane without a parachute and pretend it's you. **Before you touch the floor or ground after jumping off the step or out of an airplane, you are freefalling in a geodesic orbit of the Earth's mass and energy.** You will continue to accelerate and freefall unimpeded in your orbit, until alas, your freefall geodesic is impeded as you hit the floor or ground. (Ouch!)

But get the Earth's surface out of the way by drilling a tunnel through the Earth for you to fall into. Now you would freefall through the Earth and come out the other side. Your geodesic freefall orbit would decelerate you and you would slow and fall back to and through the Earth. You would continue in this orbit within the Earth just like the Moon or a spaceship continues in geodesic orbit around the Earth.

How the Higgs fields and bosons add mass (and energy) to particles and objects and the mass and energy acts to created geodesics in spacetime is described in **section 13.7.**

To better understand how gravity and other acceleration warp spacetime and create geodesic orbits, some thought experiments in **sections 22.7.1** through **22.7.4** will clarify the mechanics.

#22.7.1. FRICTIONLESS MERRY-GO-ROUND GEODESIC ON EARTH. Get on a frictionless merry-go-round on Earth. Have someone spin it. Since it is frictionless, you will go round and round forever in a free-fall geodesic curve determined by the dimensions of the merry-go-round. To stop and get off takes energy (acceleration).

#22.7.2. FRICTIONLESS MERRY-GO-ROUND GEODESIC IN OUTER SPACE. Move your frictionless merry-go-round to outer space far, far away from any source of gravity. Your free-fall geodesic orbit will be the same. You will go round and round forever in a free-fall geodesic curve determined by the dimensions of the merry-go-round. To stop and get off takes energy (acceleration).

Did you notice as the merry-go-round turned it wobbled? Hopefully, you built a very, very heavy merry-go-round to minimize this wobble. As the moon goes around the Earth and the Earth goes around the Sun this tiny wobble also occurs.

#22.7.3. WEIGHT ON STRING IN GEODESIC ORBIT IN OUTER SPACE. While you are in outer space, modify your merry-go-round by attaching a heavy weight on a strong string to the frictionless rotating center of the merry-go-round. Use a crank at the center of the merry-go-round to spin it fast enough that the heavy weight whirls around and around.

Now stop cranking and notice that the weight keeps spinning and spinning (with a slight wobble) around the frictionless center of the merry-go-round in a geodesic path at a velocity determined by the length of the string and the mass of the heavy weight.

#22.7.4. WEIGHT IN GEODESIC AROUND MASSIVE OBJECT. Now replace the string on the heavy weight with the warpage of spacetime (geodesic path) caused by the acceleration between the heavy object and massive object. The heavy object will continue to follow the geodesic curve orbit the massive object (with a slight wobble) such as a planet orbiting the Sun or the Moon orbiting the Earth. This orbiting will continue until friction causes them to collide.

#22.8. METRICS ARE VERY IMPORTANT PART OF GENERAL RELATIVITY.

Metrics are a vital part of mathematics. More complex metrics are required in general relativity than in special relativity. General relativity deals with the geodesic curves of space that arise due to mass, energy, and acceleration. Some examples from special and general relativity are given below to give an overview of this subject.

ONE-DIMENSIONAL METRIC. One problem in mathematics is to translate from the coordinate system of one observer to the coordinate system of another observer.

Assume a line that runs from minus **infinity** to plus infinity. The line is calibrated in **meters.** One observer's coordinate system is at zero (0). Another observer's coordinate system is zero at 10 meters.

Two observers O and O' want to know the distance D and D' between two points on the line. The first observer calls the two points $d_1 = 3$ and $d_2 = 7$.
For first observer, $D = d_2 - d_1 = 7 - 3 = \mathbf{4}$.
To translate from the first observer's coordinates to the second observer's: $D' = D - 10$
$$(d_2 - 10) - (d_1 - 10) = (7-10) - (3-10) = -3 - (-7) = -3 + 7 = \mathbf{4}$$

The key point here is that the two observers will differ greatly where the two points are, but they will agree on the distance between the two points: $D = D'$. D is called an invariant. It wouldn't matter where observers are located, all observers will measure the distance between the same two points to be the same as all other observers.

Invariants and the mathematics of invariants (scalars, vectors, and tensors) form the basis of Einstein's General Relativity.

#22.9. TRANSITIONING FROM SPECIAL TO GENERAL RELATIVITY.

The Lorentz equations in **section 21.6** are an example of using an invariant S to simplify the solution of equations.

The Invariant S is an invariant or "fact" of the Universe. It is invariant no matter what coordinate system is used to specify it, although it might need to be eventually translated to a coordinate system preferred by the user. More generally, an invariant is a quantity that is conserved in an interaction such as momentum, energy, electric charge, or specific event at a specific place at a specific time.

This equation for S^2 was shown in **section 21** to lead to the Lorentz equations for relative velocities, distances, and times for two observers moving at a constant velocity with respect to each other. The Lorentz equations for relative motion along the x/x′ axis only is:

$$x' = [\,(x) - (vt)]\,[1 - (v/c)^2]^{-1/2} \qquad x = [\,(x) + (vt')]\,[1 - (v/c)^2]^{-1/2}$$

$$ct' = [(ct) - (vx/c)]\,[1 - (v/c)^2]^{-1/2} \quad ct = [(ct) + (vx'/c)]\,[1 - (v/c)^2]^{-1/2}$$

Two different events are written as follows, when substituting ct = iT, where i = $\sqrt{-1}$ (square root of -1)

$$(x_2 - x_1)^2 + (y_2 - y_1)^2 + (z_2 - z_1)^2 + (T_2 - T_1)^2 = (S_2 - S1)^2$$

$$= (x'_2 - x'_1)^2 + (y'_2 - y'_1)^2 + (z'_2 - z'_1)^2 + (T'_2 - T'_1)^2$$

These equations make it clear that we live in a four-dimensional Universe of interrelated space (x, y, z) and spacetime (T). These equations show that spacetime is just another dimension stretching out not only forever in the past but also stretching out forever in the future just as do our normal three spatial dimensions stretch out in all directions forever. This implies that along x, y, z and T axes lie the entire past from everywhere, the entire present from everywhere, and the entire future from everywhere.

The future exists along these axes just as do the past and present. (The past and future may exist somewhere, but may not except in limited cases be accessible.) The past, present, and future are perceived differently for different observers at different locations and times.

To an observer watching a supernova that exploded millions of years ago in a telescope, the past of the supernova is very real and is happening in the observer's present and future. The past of the supernova always exists. Likewise, any specific moment of the future of the supernova always exists somewhere (far away) as well - the astronomer watches it unfold.

Certainly, you are aware of your present before anyone else; you are the first to know about your present. Your present can only be found out elsewhere no faster than the speed of light. A rocket ship could leave now from a distant star and travel to your home address and arrive long in your future. As they travel at very near the speed of light, the space traveler's onboard clocks may have advanced only a few years while your clocks and future have advanced hundreds of years.

Einstein realized that in formulating General Relativity, he should identify the **invariants** and find a way to **specify them in general coordinates** that are **easy to work in** and **easily translatable to other coordinate systems** when appropriate. Invariants are quantities that are conserved in an interaction, such as energy, momentum, pressure, density, and (mechanical) stress and strain. Such invariants are called tensors. The mathematics of such invariants is tensor analysis. Tensor analysis greatly simplifies the manipulation of equations, changing almost impossible tasks of formulating the equations of General Relativity into easily managed tasks in many instances.

#22.10. GENERALIZING COORDINATES.

Our equation for constant velocity (special) relativistic motion is:

$$x^2 + y^2 + z^2 + T^2 = S^2 = x'^2 + y'^2 + z'^2 + T'^2$$

This can be written in numerous other different ways by generalizing the coordinates x, y, z, and T such as:

$$x_1^2 + x_2^2 + x_3^2 + x_4^2 = S^2 = y_1^2 + y_2^2 + y_3^2 + y_4^2$$

There is no difference in meaning of the equations so long as it is known how to translate the equations to the desired coordinate system. It is not necessary to do this until all the manipulations are finished, usually in a much easier and faster manner. Sometimes the coordinates are labeled 0, 1, 2, 3. In this case, T or -ct is usually x_0.

Here's an even more COMPACT AND POWERFUL way to write the above equation:

$$x_i^2 = S^2 = y_j^2 \quad (i, j = 1, 2, 3, 4)$$

Or, this equation can also be represented as:

$$x_i^2 = S^2 = x'_i^2 \quad (i = 1, 2, 3, 4)$$ Note that the superscript 2 means to square the quantity. It does not take any other value like the i's and j's do.

However, this equation only applies to straight lines (such as constant velocity, and does not reflect the curved paths of accelerated motion.

Einstein derived his equations for the curved motion of acceleration and mass and energy starting as follows (where d means a small change along the curved path or a coordinate of the curved path) In calculus, the d stands for derivative.):

$$ds^2 = dx_i^2 = dx_1^2 + dx_2^2 + dx_3^2 + dx_4^2 \quad \text{(Note the special meaning of } d_i \text{)}$$

To be applicable to all coordinates, a constant needs to be added as follows:

$$ds^2 = a_i dx_i^2 = a_1 dx_1^2 + a_2 dx_2^2 + a_3 dx_3^2 + a_4 dx_4^2$$

For polar coordinates of two dimensions r and θ,

$$ds^2 = a_i dx_i^2 = a_1 dx_1^2 + a_2 dx_2^2 \quad\quad\quad\quad \text{(22.3-1)}$$

$$ds^2 = dr^2 + r^2 \, d\theta$$

$$a_1 = 1 \quad\quad dx_1^2 = dr^2 \quad\quad a_2 = r^2 \quad\quad dx_2^2 = d\theta$$

This is a good start on the road to general relativity. The next step is to include the study of tensors that are necessary to form the equations of motion and acceleration for curved (accelerated) paths.

#22.11. USING TENSORS TO DESCRIBE MOTION IN CURVED SPACETIME.

A scalar is a tensor of rank 0.
A vector is a tensor of rank 1.
Tensors can also be of rank 2, 3, 4, ...

A scalar (temperature, density, speed, age, etc.) has magnitude only (one component) at a given point, but not direction
A vector (velocity, force, etc.) has both magnitude and direction (three components).
A tensor of the second rank (pressure or force acting on an area) has magnitude and direction acting on an area (9 components).
A tensor of the third rank has 16 components, etc.

Equation 22.3-1 above can now be generalized to any curved surface as follows:

$$ds^2 = g_{uv} (dx_u \cdot dx_v)$$

For two dimensions has four terms:

$$ds^2 = g_{11} \cdot dx_1{}^2 + g_{12} \cdot dx_1 \cdot dx_2 + g_{21} \cdot dx_2 \, dx_1 + g_{22} \cdot dx_2{}^2$$

Three dimensions would have nine terms; four dimensions would have 16 terms.

Einstein's equations employ tensor calculus. In solving these equations, Christoffel Symbols are used to compactly formulate his equations

#22.12. EINSTEIN'S GENERAL RELATIVITY FIELD EQUATIONS.
[Condon and Odishaw, p. 2-50, 2-51]

Einstein's equations have proven to be highly accurate and in close agreement with experimental results - much more so than Newton's equations.

The **Bibliography** lists several texts on General Relatively for further information and study. (See **section 22.2** for additional information about the mathematics of General Relativity.) Several different approaches in presenting Einstein's equations are used by these references.

Einstein's field equations stated in compact form are:

$$E^{\lambda\mu} = - kT^{\lambda\mu} \quad cm^{-2} \qquad (22.5.1\text{-}1)$$

Don't let the simplicity of these equations misled.   Writing them out completely and lowering the tensor indices, Einstein's field equations take the form:

$$R_{\lambda\mu} - \tfrac{1}{2}\, g_{\lambda\mu}\,(R - 2\,C) \; = \; - kT_{\lambda\mu} \qquad cm^{-2} \quad (22.5.1\text{-}2)$$

Where:

C is the cosmological constant. (The symbol Λ is
sometimes used instead of C.)

c is the velocity of light

E is the distribution of momentum and energy representing the
distribution of matter and radiation in the Universe.

$G = 6.670 \times 10^{-8}$ dyne-cm/g^2 (Newton's gravitational constant)

$g_{\lambda\mu}$ is the spacetime metric of the Universe

$k \; = \; 8\pi G/c^2 = 1.864 \times 10^{-27}$ cm/g

R is scalar curvature

$R_{\lambda\mu}$ is   the Ricci curvature tensor

$T_{\lambda\mu}$ is stress-energy momentum density.

The left side of this equation determines the geodesic path any object will take when moving in curved spacetime of the Universe. The right side of the equation is the amount and density of mass-energy-stress-momentum-pressure that determines the amount and location of curvature (warp) of spacetime.

Einstein first added the cosmological constant C to his equations and then decided that the Universe was static so the value of C should be zero. When Hubble found that the Universe was expanding, Einstein realized his great mistake as he would have predicted this phenomenon and it would have been a tremendous scientific theoretical achievement in its own right.

The earliest and most important exact solution of Einstein's Field Equations was by Karl Schwarzschild in 1916. Einstein was surprised that a solution to his equations was found so soon. Schwarzschild's solution also predicted what are now called black holes. Schwarzschild's solution led directly to calculating the perihelion of mercury. See the relativity references in the **Bibliography** for further information about Schwarzschild's solution and other solutions to Einstein's equations.

Einstein's General Relativity Field equations expand to 16 equations for the 16 values that $T_{\lambda\mu}$ can take on with $\lambda\mu$= 11, 12, 13, 14, 21, ... , 24, ... , 44. These 16 equations quickly reduce to 10. These 10 are interrelated and very difficult to work with and solve. However, with the assumption of a universe that is uniform in distribution of mass and energy, the 10 can be reduced to one. This type of assumption may not be true for a small sample of the Universe, but over a very large sample it is remarkably correct. Such simplifying assumptions are very common in physics and enable solution of many otherwise intractable problems.

#22.13. COMPARISON OF EINSTEIN'S EQUATIONS WITH NEWTON'S EQUATIONS.

As has been pointed out earlier, Newton's equations, in many if not most situations, provide extremely accurate results and are adequate. However, physicists cannot rest if experimental results and calculated results do not coincide at least within expected experimental tolerances. As experimental techniques improved, the (although) small variances in Newtons's equations fell out of expected experimental tolerances. In such cases, physicists (and others) found this intolerable and looked for the source of the errors and whether due to the mathematical technique and theory or the experimental technique. This dilemma led to Einstein developing his General Theory of Relativity.

The following paragraphs compare Newton's equations with the more accurate Einstein General Relativity equations (**section 22.12.**) The references cited in **section 22.3** provide detailed derivations.

#22.13.1. PLANET MOVING AROUND THE SUN - THE PERIHELION OF MERCURY.

Newton's equations for Mercury moving around the Sun maps out an ellipse. Newton's ellipse does not rotate (have a perihelion).

Careful experiments by astronomers found that Mercury's orbit maps out an ellipse for which the perihelion (the closest point of the ellipse to Mercury) moves around Mercury at 43 seconds of arc per 100 years. This is a very, very small tiny (1 degree = 60 minutes = 3600 seconds). Nevertheless, physicists and astronomers were horrified as it showed that Newton's equations did not accurately describe the actual situation.

NEWTON'S EQUATIONS FOR PERIHELION OF MERCURY:

$$d^2u/d\phi^2 + u = m/h^2 \qquad \text{(d represents the derivative)}$$

$$r^2 \, d\phi/dt = h$$

r = distance from sun to planet.
u = 1/r
h = constant
m = mass of sun.
ϕ = angle swept out by planet in time t.

When Einstein developed his General Relativity equations (**section 22.12**) in 1915, one of his first applications was to use them to compute the orbit of Mercury around the Sun [Lieber p. 270]. Comparing with Newton's equations shows Einstein's equations add the term $3mu^2$.

EINSTEIN'S EQUATIONS: FOR PERIHELION OF MERCURY:

$$d^2u/d\phi^2 + u = m/h^2 + 3mu^2$$

$$r^2 \, d\phi/dt = h$$

When Einstein calculated the perihelion of Mercury, he obtained the value of 43 seconds of arc per 100 years, well within the experimental results. Einstein was so elated that it took days for him to regain his composure!!!

#22.13.2. DEFLECTION OF LIGHT RAY BY MASSIVE OBJECT.

It seems fitting that the final section in this book touches on both Isaac Newton (**section 1.2.3**) and Albert Einstein (**section 22**). Since the times of Johann Kepler (1571 -1630) and Isaac Newton (1642-1727), their insight about motion and gravity has been a mainstay of science. Their mathematics for the paths of celestial objects and the laws of gravitational attraction are still more than adequate and a mainstay in many situations.

Later physicists and astronomers expanded and refined the work of Kepler and Newton to develop many complex mathematical techniques to solve and predict the complex motion of planets, moons, suns, other celestial objects, satellites, bombs, artillery shells, and even the motion of electrons around the nucleus of an atom.

In review, the gravitational Force of attraction F_g between two objects such as the sun M_S and the Earth M_E (**sections 1.2.3 and 2.3.3**):

$F_g = (G_N\ M_S\ M_E)/r^2$ where:
F_g Gravitational force between the objects in Newtons (N)   (1 N = 1 kg m/sec^2)
M_S = mass of Sun = 1.998 x 10^{30} kg
M_E = mass of Earth = 5.972 x 10^{24} kg
M_p =mass of photon = 2.2 x 10^{-28} kg   (**section 5.4**)
r = distance between the centers of the two masses in meters as they pass
      one another
G_N = gravitational constant = 6.67385 x 10^{-11} N m^2 /kg^{-2} (m^3 kg^{-1} s^{-2})

The acceleration of all objects due to the gravity of another object is the same (**section 1.2.3, section 2.5.3**)

D = Distance Sun to Earth = 1.496 x 10^{11} meters
R_E = Radius of Earth = 6.378 x 10^6 meters
R_S = Radius of Sun   = 6.9551 x 10^8 meters

L = Length of light ray from outer edge of Sun to Earth = 1.50297 x 10^{11} meters
C = Velocity of light = 2.9979 x10^8 meters/sed
Time for light to travel between Earth and Sun = 499.016 sec = 8.3169 min.
Time for light to travel from outer edge of Sun to Earth = 501.35 sec = 8.356 min.
Angle from outer edge of sun to Earth = 0.265137 deg.

Putting this all together and solving the three body equation for the Earth and Sun and a photon traveling past the Sun to the Earth is a complex problem. There are many simplifying assumptions that last-century astronomers would make such as reducing a three body problem to a two body problem (**section 1.2.3**). The results find the bending of the light ray to be about 0.8 seconds of arc.

However, the above results before the publishing of Einstein's General Relativity were about 1/2 of the actual experimental value because, among other things, the velocity of light is slowed as it passes a massive object such as the sun, giving the light rays more time to be deflected. Einstein's equations give accurate results as follows [Collier p. 266]:

$$\alpha = 4GM/c^2b$$

Where:
α = angle of deflection of light ray just grazing radius r of Sun
M = mass of sun = 1.99 x 10^{30} kg
G = gravitational constant = 6.67 x 10^{-11} N m^2 k^{-2}
b = distance r to center of sun at closest approach = 6.96 x 10^8 m

Putting these values into the equation above derived from Einstein's general Relativity Equation, gives an α of 1.75 seconds of arc in agreement with experimental results. Since first published, Einstein's General Relativity equations have been used extensively and continue to provide results that accurately reflect how our Universe works.

The techniques of Kepler, Newton, and astronomers who built on their genius are still used today in many, many situations where relativistic effects are not significant. However where relativistic effects are large at high velocities and/or extreme gravitational forces, then the mathematics of General Relativity are required.

#EPILOGUE.

With the efforts going on in the fields of particle physics, relativity, cosmology, astronomy, and string theory, any work such as this covering broad and important issues of the physics of our Universe and Space Beyond, is bound to be at best a work in process. Nobel Prize winning breakthroughs may occur at any time. New theoretical elementary particles such as the Higgs boson may be found experimentally at any time, despite the eminent physicists that bet against their existence. As the characteristics of the Higgs boson are studied, understanding of the Higgs fields and gravity may be radically changed.

One of several hypotheses described herein is that Spacetime beyond our Universe will be found to be just like the seething empty spacetime in our Universe EXCEPT MUCH, MUCH COLDER. Time marches on there just like it exists and marches on in empty space here. Spacetime beyond our Universe will be filled with a background of virtual particles and perhaps several more universes of various kinds that look similar to our Universe separated at least in our case by great distances.

However, until recently, many if not most physicists and cosmologists believed otherwise - that there was nothing beyond our Universe and time did not exist before our Universe. Their position was that spacetime is unfurling like a balloon or "bubble" as our Universe expands at an accelerated rate. However, nothing not even time existed before our Universe, and doesn't exist anywhere else. (My disagreement prompted me to write the first edition of this book in 2011.) Now, this situation has changed and most physicists believe that there was time and space before the Big Bang and will be long after this Universe has disappeared.

This is an experimental question. To verify the mathematical hypothesis (conjecture), the experiment will require a warp-speed space craft to travel beyond our Universe, explore what is there, and report back. Unfortunately, no such space craft will exist for perhaps fifty or one hundred years, if ever. Our current time table seems to be to put people on Mars in about 10 to 20 years, put people on a hospitable planet in another solar system in fifty years, and then maybe traveling beyond our Universe in 100 years.

To a great extent, current opinion is based on the success of the inflationary expansion theory of the Universe during the Big Bang. This inflationary expansion had to take place at a pace that for a short time was **greater than the velocity of light** – an "impossibility"! To save the inflation theory, it was postulated that the Universe must be "unfurling" like a balloon or bubble into "nothing" – no time, no space, no nothing. This was deemed to be okay, and not a violation as it was not in an inertial system. The inflationary theory was vindicated. A college industry of bubble and membrane universes grew up accordingly.

In writing this edition as prior editions of this book, it seemed (to the writer) that throwing out infinite empty space just went too far. Doing so also threw out all that could go on there such as the uncertainty principle, virtual particles, Higgs fields, probability waves, time, and the like; and most important of all, a place for the Big Bang itself to spend its youth. Accordingly an alternative was theorized herein as a rescue of infinite empty space beyond our Universe. It was theorized that at temperature of much less than 1 degree K (very near absolute zero degrees Kelvin, the Higgs fields might undergo a change in state, were frozen (like ice perhaps), and did not impede velocities or limit them to the speed of light or less.

At the Big Bang then, the inflationary expansion could exceed the speed of light until the "frozen" Higgs fields thawed out and began to function normally. The uncertainty principle, virtual particles, Higgs fields, probability waves, time, and the like were safe in infinite empty space long before, before, during, and after the Big Bang. The Big Bang even had a place to call home during its youth before going Bang! At the temperature almost zero Kelvin before the Big Bang, virtual particles could form and provide the fuel of a future Big Bang. Although at almost zero K, it would take a long, long time for a Big Bang to develop. Finally inter-universe travel outside our Universe to other universes could take place at unlimited and unhindered velocities.

There are many mathematical models of the beginning, current status, and future of our Universe [Collier chapter 11]. The model that is thought to best describe our Universe is the "**Accelerating Model**" as our Universe underwent an accelerating inflationary expansion at its inception (50,000 years), then slowed (9.8 billion years), and is now expanding at an accelerating rate due to Dark Energy (4 billion years). This model also predicts the age of the Universe to be 13.8 billion years which is the current estimate (**section 6.6**).

The **Accelerating Model** of the Universe of the Universe is also known as the Lamda-CDM Model where CDM stands for Cold Dark Matter (**section 14.2.1).**

It has been a theme early in (**section 2.10.3**) and throughout this book (i.e. **section 14.4.2**) that Dark Energy is the negative energy vacuum of empty space beyond our Universe. Whether this theory proves valid or not remains to be determined.

There are other areas throughout this material that are subject to the great argument settler of experiment. Life is interesting and exciting exploring particle physics, cosmology, relativity, and string theory while riding on rotating Earth. As Earth rushes through space revolving around the sun, it seems as if there is a big surprise with every revolution and sometimes with every rotation. Sooner or later, repeatable experiments will have their way with me and everyone else. That's why this book was written.

JGB
Send Comments **to: JBeanMBean@msn.com**

#BIBLIOGRAPHY.

The material listed below was reviewed and is recommended for further reading particularly from a historical viewpoint which was essentially ignored herein. Much material is similar between references and it is difficult to single out and cite any one reference in any specific area except a provided by the titles or as noted below. In cases of conflict between sources the *Review of Particle Physics* (*RPP*) prevailed. Citations in the text to **BIBLIOGRAPHY** listings are enclosed in brackets [   ].

[1] Baggott, Jim, *The Quantum Story*, Oxford University Press (2011).

[2] Baggott, Jim, *Higgs*, Oxford University Press (2012).
A historical summary of the search for the Higgs boson by a participant.

Beringer, J. et al, *Review of Particle Physics* (Particle Data Group), Physical Review D, Vol. 86 010001 A 1525 page definitive summary of the latest research and experimentation in particle physics. An abbreviated pocket booklet is also published; order it at: http://pdg.1b1.gov (2012). See also listing for Olive, K.A.

Bohm, David, *Quantum Theory*, Dover Publications, Inc. (1951).
A qualitative and physical presentation of fundamentals followed by considerable mathematical detail.

Collier, Peter, *A Most Incomprehensive Thing: Notes Toward a Very Gentle Introduction to the Mathematics of Relativity* (2012). (Available at Amazon.com)

Condon, E. U. and Odishaw, Hugh, *Handbook of Physics*, McGraw Hill (1967).

Freeman, Morgan, "Through the Wormhole (Video), "Is Gravity an illusion?" Science Channel.

Feynman, Richard P., *QED the Strange Theory of Light and Matter*, Princeton University Press (2006).
Richard Feynman is developer of Feynman diagrams of particle interactions.

Gilmore, Robert, Lie Groups, Lie Algebras, and Some of Their Applications, Dover(2002).

[1] Greene, Brian, *The Hidden Reality*, Vintage Books (2011).

[2] Greene, Brian, *The Fabric of the Cosmos*, Vintage Books (2004).

[3] Greene, Brian, *The Elegant Universe,* Vintage Books(2000).

Griffiths, David, *Introduction to Elementary Particles*, Wiley-VCH (2010).
A textbook on the mathematics of Particle Physics. Chapter 10 "Gauge Theory) gives the mathematical derivation of the Higgs boson from the Lagrangian.

[1] Hawking, Stephen, *A Brief History of Time* (2004).
[2] Hawking, Stephen and Mlodinlow, Leonard, *THE GRAND DESIGN* (2008).
[3]Hawking, Stephen, *Curiosity – Did God Create the Universe,* Discovery Productions, DiscSc Channel, (Video recorded 8/7/2011).

Krass, Lawrence M. *A, Universe from Nothing*, Free Press (2012).

Kay, David C, *Tensor Calculus,* McGraw-Hill (1988

Lieber, Lillian R, *The Einstein Theory of Relativity,* Paul Dry Books (2008)
General Relativity using tensor notation and calculus.

Lipschutz, Seymour and Lipson, Marc, *Linear Algebra,* McGraw-Hill (2002).

[1] McMahon, David, *Relativity*, McGraw-Hill (2006).
[2] McMahon, David, *String Theory*, McGraw-Hill (2009)

Olive, K. A. et al, *Review of Particle Physics* (Particle Data Group) (2014)
Palen, Stacy, *Astronomy, McGraw*-Hill (2002), D. Van Nostrand Co., Inc. (1955)

Riordan, Michael and Zajc, William A, *The First Few Microseconds*, Scientific American (May 2006)

Robertson, John, *Geometrical and Physical Optics,* D. Van Nostrand Co., Inc. (1955)
Schumm, Bruce A, *Deep Down Things,* The John Hopkins University Press (2004).

Stannard, Russell, Relativity*, a Very Short Introduction,* Oxford University Press (2008).
Sternheim, Morton M, and Kane, Joseph W, G*eneral Physics,* John Wiley and Sons (1986)

#INDEX

Topic   **_Section Number_**

ACCELERATION AND DECELERATION OF TRAVELER'S CLOCK TAKE TRAVELER'S CLOCK
     (AND TRAVELER WITH IT) TO NEW (SLOWER) CLOCK RATE 21.12.4

ACCELERATION BY GRAVITY   22.4

ADVANCED PARTICLE SEARCHES AND TOPICS 14

AGE AND DIAMETER OF UNIVERSE 1.2.14

ANGSTROM 1.2.12

ANGULAR MOMENTUM 1.2.8

ANTI-PARTICLE AND PARTICLE UNIVERSES   15

ANTI-PARTICLE INTERACTIONS 13.2

BARYONS - DESCRIPTION 11.7

BARYONS, EXAMPLE 10.3

BARYONS, LISTINGS 10.4

BARYONS, PROTONS AND NEUTRONS ARE MADE OF THREE QUARKS 11.7.2

BASIC PHYSICS OF THE EARTH AND UNIVERSE 2

BEAT FREQUENCIES (NOTES) 3.11

BIBLIOGRAPHY

BIG BANG 17   18

BIG BANG, CREATED BY QUANTUM VIBRATING STRINGS   17.5

BIG BANG, AS TEMPERATURE AND PRESSURE INCREASE IN BIG BANG SPHERE,
     BIG BANG OCCURS   17.7

BIG BANG, SPHERE OF LEPTONS, QUARKS, GLUONS, AND BARYONS NOT ABLE TO
     OBTAIN HIGH ENOUGH PRESSURE, TEMPERATURE, AND DENSITY TO CAUSE
     17.3

BIG BANG, HOW MUCH ENERGY DOES IT TAKE TO MAKE OUR UNIVERSE?   17.7

BIG BANG, WHERE AND HOW DID COME ABOUT BEFORE OUR UNIVERSE EXISTED? 17

BIG BANG, VIBRATING STRING CHARACTERISTICS INITIATED WITH LESS PRESSURE,
     TEMPERATURE, AND ENERGY THAN EXPECTED   17.8

BIG BANG, AS COOLED, HYDROGEN WAS FORMED   6.2 6.2.1

BIG BANG: LONG BEFORE AND LONG AFTER   18

BIG BANG, TIMELINE, TEMPERATURE, AND PRODUCTS 6.2

BINOMIAL THEOREM   12.2.1

BLACKBODY RADIATION: ORIGIN OF QUANTUM THEORY OF LIGHT 4.2

BOSONS 9.4

BOSONS – DESCRIPTION & RELATIVE STRENGTH 11.4

CARTESIAN ORTHOGONAL COORDINATES OF LORENTZ EQUATIONS 21.7

CENTER OF MASS   1.2.3

CHARGE OF ELECTRON AND PROTON   1.2.15

CHARGES, ELECTROMAGNETIC, OF ELEMENTARY PARTICLES ARE INFINITE 13.3.1

CHANGE OF STATE   2.9.4

CMBR 6.7

COHERENT WAVES 3.10

 COHERENT WAVES INTERFERE WITH EACH OTHER 7.1

COMPOSITE PARTICLES - HADRONS (MESONS, BARYONS, TETRA- AND
        PENTA-QUARKS) 10

CONDUCTION, TRANSFER OF HEAT ENERGY BY   2.9   2.9.2

CONSERVATION OF KEY PARTICLE CHARACTERISTICS 12.2

CONVECTION, TRANSFER OF HEAT ENERGY BY   2.9   2.9.1

COORDINATE ROTATION AND SPACETIME GRAPHS 21.8

COORDINATE TRANSFORMATION 21.9

COORDINATE TRANSFORMATION USING LINEAR CARTESIAN COORDINATES 21.6

CONSERVATIVE FORCES AND DISSIPATIVE FORCES 2.5.2

CONSTANT RELATIVE MOTION   21.2

CONSTANTS OF THE EARTH AND UNIVERSE 1.2.

COSMIC MICROWAVE BACKGROUND RADIATION (CMBR) 6.9

COSMIC RAYS 1.2.16   6.11

COSMOLOGICAL UNITS 6.1

COULOMB, COULOMB'S LAW   1.2.15   13.3

CREATION OF ELEMENTS IN THE UNIVERSE   6.2

CURVING (WARPING) OF SPACETIME   22.5

DARK ENERGY (72 PERCENT OF OUR UNIVERSE) 14.2   14.3

DARK ENERGY, CAUSING UNIVERSE TO EXPAND   14.3

DARK MATTER (21 PERCENT OF OUR UNIVERSE) 14.2.1   14.2.2

DARK MATTER, COMPOSITION AND ORIGIN   11.7.4   14.2.2

de BROGLIE WAVE CHARACTERISTICS 1.2.20   12.3

INDEX (CONTINUED)

DEFINITIONS 1.2

DEFLECTION OF PHOTONS BY GRAVITY 5.5

DIAMETER, CURRENT, OF UNIVERSE 1.2.14   6.1   6.8

DIAMETER OF PROTON   1.2.18

DIAMETER OF QUANTUM OF ENERGY Q_e (STRING) 1.2.19

DIAMETER OF UNIVERSE BEFORE THE BIG BANG 1.2.14   6.1

DISSIPATIVE FORCES 2.5.2

DOPPLER EFFECT   3.6.1

EARTH, LIFE, AND SCIENCE TIMELINE 1.1

EINSTEIN'S EQUATIONS 22.12

EINSTEIN'S EQUATION OF EQUIVALENCE OF MASS m AND ENERGY E 1.2.1

E EINSTEIN'S FIELD EQUATIONS 22.12

EINSTEIN'S UNIFORM UNIVERSE EQUATIONS 22.12

ELEMENTARY PARTICLE DESCRIPTIONS 11

ELEMENTARY PARTICLES – FERMIONS (LEPTONS AND QUARKS) & BOSONS 9

ELECTROMAGNETIC CHARGES OF ELEMENTARY PARTICLES ARE INFINITE 13.3.1

ELECTROMAGNETIC FORCE BOSON (PHOTON) 9.4.1.

ELECTROMAGNETIC FORCE BOSONS (PHOTONS) MEDIATE INTERACTIONS 13.3

ELECTROMAGNETIC RADIATION AND PHOTONS 5

ELECTROMAGNETIC SPECTRUM 1.2.12   5.2

ELECTROMAGNETICALLY CHARGED LEPTONS 9.2   11.2.2

ELECTROMAGNETICALLY NEUTRAL LEPTONS (NEUTRINOS) 9.1   11.2.1

ELECTROMAGNETIC WAVE VELOCITY   3.5.2

ELECTROMAGNETIC WAVES PROPAGATE FROM CENTRAL POINT   21.1.4

ELECTRON VOLT Ev   1.2.15   12.1

ELEMENTS IN THE UNIVERSE   6.2 6.2.1 6.2.2   6.2.3

EMPTY SPACE, ENERGY BORROWED FROM, TO CREATE UNIVERSE MUST BE REPAID 17.9

EMPTY SPACE IS NOT EMPTY   12.7   12.7.1   12.7.2

EMPTY SPACE SCREENS TRUE VALUE OF PARTICLE ELECTRIC CHARGE 13.3.2

EMPTY SPACE, FABRIC THAT   EXISTED BEFORE BIG BANG CREATED VIRTUAL PARTICLES
     AND VIRTUAL ANTI-PARTICLES THAT LED TO SPHERE OF BIG BANG   17.2
ENERGY   2.5 2.5012.5.2 2.5.3   2.9   12.2
ENERGY CONVERSION LIST 1.2.5 1.2.6   1.2.25
ENERGY (E) OF PHOTON 1.2.10   5.3
ENERGY, KINETIC   1.2.6   2.1   2.5.1   2.5.3
ENERGY, POTENTIAL, CONSERVATIVE FORCES, AND DISSIPATIVE FORCES 2.5.2   2.5.3
ENERGY WARPS SPACETIME   21.1
ENTANGLED PARTICLES 12.6
EPILOGUE
EQUATIONS OF PHYSICS MUST BE MODIFIED FOR SPECIAL RELATIVITY EFFECTS 21.13
EQUIVALENT MASS OF PHOTONS 1.2.11   5.4
ESCAPE VELOCITY   2.5.3
ETAC, EXPERIMENTALLY TO A CERTAINTY 9
EXAMPLE BARYONS 10.3
EXAMPLE MASONS 10.1   10.2   11.6

FERMIONS – DESCRIPTION 11.1
FERMIONS (LEPTONS AND QUARKS) AND BOSONS 9
FERMIONS (LEPTONS AND QUARKS) INTERACT WITH WEAK CHARGE 13.5.1
FEYNMAN DIAGRAMS 13.1.2
FIELDS, MATHEMATICAL DESCRIPTION OF 13.7.3
FINE STRUCTURE 4.0
FORCE 1.2.2, 2.3.1, 2.3.4
FORCES, RELATIVE STRENGTH   11.4
FORCE, UNITS OF 2.3.4
FOURTH DIMENSION, IS "T" or "t"? 21.11
FREQUENCY 1.2.9   3.6
FREQUENCY, VELOCITY, AND WAVELENGTH OF LIGHT 1.2.9   5.1

GENERAL RELATIVITY   21.1   22   22.12
GENERAL RELATIVITY EQUATIONS   22.12
GENERAL RELATIVITY, QUALITATIVE ASPECTS OF   22.3
GENERAL RELATIVITY, QUANTITATIVE ASPECTS OF 22.2

INDEX (CONTINUED)

GENERAL RELATIVITY, TRANSITIONING FROM SPECIAL 22.9

GENERALIZING COORDINATES 22.10

GEODESIC ORBITS   22.7

GRAVITATIONAL FIELD 1.2.3

GRAVITATIONAL FORCE 1.2.3

GRAVITATIONAL WAVES 2.3.3

GRAVITATIONAL FORCE BOSONS 9.4.4

GRAVITON BOSONS   13.7

GRAVITY, ALTERNATE THEORIES   13.7.9

GRAVITY 2.3.3

GRAVITY, POSITIVE AND NEGATIVE   2.10.3

HARMONICS OF WAVES   3.6

HADRONS ARE PRODUCED WHEN QUARKS INTERACT WITH GLUONS   13.4.3

HADRONS (MESONS AND BOSONS) – DESCRIPTION   11.5

HEAT, RELATIONSHIP TO OTHER FORMS OF ENERGY   1.2.24

HEATING OBJECTS TO VARIOUS CONSTANT TEMPERATURES   4.1

HIGGS BOSON, DETECTED IN JULY 2012   13.7.10

HIGGS BOSONS AND GRAVITON BOSONS   13.7

HIGGS FIELDS   13.7.4   13.7.5

HIGGS FIELDS AND BOSONS CORRECT THEORETICAL PROBLEMS   13.7.4

HIGGS FIELDS, DESCRIPTION   13.7.5

HIGGS FIELDS IMPART GRAVITATIONAL FORCE CHARACTERISTICS   13.8

HIGGS FIELDS, TAKE ON SCALAR VALUES   13.7.6

HYDROGEN, WAS FORMED AS BIG BANG COOLED   6.2.1.

HYPERBOLIC FUNCTIONS OF LORENTZ EQUATIONS 21.5

HYPERFINE SPLITTING 4.0

INTERFERENCE - WAVES CANCEL AND REINFORCE EACH OTHER AND THEMSELVES   3.9

INTER-UNIVERSE TRAVEL   Epilogue

KINETIC ENERGY   1.2.6

KINETIC ENERGY WARPS SPACETIME COLLINEARLY   21.1.1

LAGRANGIAN FORMULATION OF ENERGY AND MOTION   13.7.4

LAMB EFFECT   4.0

LENGTH CONTRACTION   21.3   21.3.1

LENGTH CONTRACTION AND TIME DILATION   21.3   21.3.1   21.3.2

LEPTON AND GENERAL PARTICLE CHARACTERISTICS   8.1

LEPTONS, DESCRIPTION   11.2

LIFE, EARTH, AND SCIENCE TIMELINE   1.1

LIGHT BEHAVIOR IS DIFFERENT THAN THAT OF OTHER TYPES OF WAVES. 21.1.3

LIGHT CHARACTERISTICS, WAVE MOTION AND   3

LIGHT, POLARIZED 3.12

LIGHT RADIATION AND SPECTRUM   4

LIGHT, VELOCITY, FREQUENCY, AND WAVELENGTH OF   1.2.9   3.5.2   5.1

LORENTZ EQUATIONS   21.2

LORENTZ EQUATIONS LEAD TO EQUATIONS FOR PROPAGATION OF LIGHT 21.4

MASS AND CHARGE OF ELECTRON AND PROTON   1.2.15

MASS AND ENERGY   2.1   22.3.1

MASS AND ENERGY CAUSE SPACE5TIME TO CURVE (WARP)   22.3.2

MASS AND FORCE   2.3.4

MASS AND GRAVITY   13.7.1

MASS AND ENERGY ARE QUANTA OF THE SAME THING   2.2

MASS, CENTER OF   1.2.3

MASS, EQUIVALENT, OF PHOTONS   1.2.11   5.4

MASS, HOW CHARACTERISTICS ARE DEVELOPED   13.7

MASS, HOW GIVEN TO PARTICLE OR OBJECT   13.7.1   13.7.5

MASS (IN KILOGRAMS) OF VARIOUS OBJECTS   1.2.4

MASS OF ELECTRON AND PROTON   1.2.15

MASS, UNITS OF   2.3.4

MASS, WHERE DID, OF NEUTRON AND PROTON COME FROM   13.4.5

MESONS ARE COLOR AND CHARGE NEUTRAL   11.6.1

INDEX (CONTINUED)

MESONS, EXAMPLE   10.1

MESON LISTINGS   10.2

METRICS IN GENERAL RELATIVITY   22.8

MILKY WAY   1.12.13

MOMENTUM, LINEAR AND ANGULAR   1.2.8   2.7   2.8

MOTION IN CURVED SPACETIME   22.5

NEWTON'S LAWS OF MOTION   2.3.5.

OSCILLATING SINUSOIDAL WAVES TRAVEL OUTWARD FROM THEIR SOURCE 3.1

PARTICLE AND ANTI-PARTICLE INTERACTIONS   13.2

PARTICLE CHARACTERISTICS   8

PARTICLE CHARACTERISTICS, LEPTON AND GENERA L 8.1

PARTICLE CHARACTERISTICS, QUARKS, BOSONS, AND HADRONS   8.2

PARTICLE INTERACTIONS   13.1   13.2

PARTICLE INTERACTIONS AND DECAY RULES   13.6

PARTICLE INTERACTIONS, BACKGROUND INFORMATION   12

PARTICLE INTERACTIONS, NOTATION   13.1.1

PARTICLES, QUANTUM, ARE SEEMINGLY AT TWO (OR MORE) PLACES AT ONCE   7.2

PARTICLES QUANTUM, INTERFERE WITH THEMSELVES   7.2

PENTA- AND TETRA-QUARKS   10.1   11.7.3

PERIODIC WAVES AND BEAT FREQUENCIES (NOTES)   3.11

PHOTOELECTRIC EFFECT   4.3

PHOTON, DEFLECTION OF, BY GRAVITY   5.5

PHOTON, ENERGY   1.2.10   5.3

PHOTON, INTERACTION   13.3.3.

PHOTON J SPIN ENABLES ELECTRIC CHARGES TO ATTRACT OR REPEL EACH OTHER 13.3.3

PHOTONS   5

 PHOTONS AND ELECTROMAGNETIC RADIATION   5

PHOTONS MEDIATE ELECTROMAGNETIC INTERACTIONS   13.3

PHYSICS OF EARTH AND UNIVERSE   2
POTENTIAL ENERGY WARPS (CURVES) SPACE AND TIME (SPACETIME)   21.1.2
PROTON, DIAMETER OF   12.18

PI π 1.2.21
PLANCK MEASUREMENTS 1.2.17
POSSIBLE PATHS OF PARTICLE 12.4.3
POWER   1.2.7   2.6
PREFACE
PERIODIC WAVES AND BEAT FREQUENCIES (NOTES) 3.11
POLARIZED LIGHT   3.12
PRESSURE   1.2.22   2.10
PRESSURE, EXAMPLES   2.10.2
PRESSURE, MEASUREMENT   2.10.1
PRESSURE, POSITIVE AND NEGATIVE 2.10.3
PROBABILITIES 12.4.2   5.5.1
PROBABILITY WAVES   12.4.2
PROPAGATION OF LIGHT   21.4
PROTON MASS 11.7.2

RELATIVE MOTION   21.2
QUARKS 9.3   11.3   11.7.3
QUARK COMBINATIONS IN BARYONS   11.7.3
QUARK COMBINATIONS IN MESONS   11.6.2
QUARKS INTERACT WITH STRONG NUCLEAR (COLOR) FORCE GLUONS 13.4.2
QUANTUM, VIBRATING, OF ENERGY Q_e (STRING) 1.2.19
QUANTUM MECHANICS, ORIGIN OF, BLACKBODY RADIATION 4.2
 QUANTUM PARTICLES ARE SEEMINGLY AT TWO (OR MORE) PLACES AT ONCE 7.2
QUANTUM PARTICLES INTERFERE WITH THEMSELVES 7.2
QUANTUM STRINGS OF VIBRATING ENERGY   17.1
QUANTUM THEORY OF LIGHT, ORIGIN OF, BLACKBODY RADIATION: 4.2
QUASARS, BLACK HOLES AND   6.3

INDEX (CONTINUED)

RADIATION, TRANSFER OF HEAT ENERGY BY   2.9   2.9.3

REFLECTION OF WAVES 3.7

REFRACTION OF WAVES 3.8

RELATIVE MOTION, CONSTANT 21.2

RELATIVISTIC MASS AND ENERGY ARE CONSERVED 12.2.1

SCIENCE, EARTH, LIFE TIMELINE 1.1

SCIENTIFIC METHOD   19

SIMULTANEITY: WHEN WAS THE SHOT FIRED IN NEW YORK CITY? 21.10

SOUND VELOCITY   **3**.5.1

SPECIAL RELATIVITY, MATHEMATICS 21.3

SPECIAL RELATIVITY, THEORY   21

SPECIAL RELATIVITY, THEORY AND MATHEMATICS 21.1

SPACETIME GRAPHS 21.8

SPECTRUM, ELECTROMAGNETIC 1.2.12   5.2

SPIN,J, OF PHOTON,   ENABLES ELECTRIC CHARGES TO ATTRACT OR REPEL 13.3.3

STARK EFFECT 4.0

STARS – CREATION OF ELEMENTS IN THE UNIVERSE   6.2.2   6.2.3

STARS, TYPES OF   6.2.2

STARS, DYING ARE FACTORIES CREATING ELEMENTS   6.2.3

STRINGS, QUANTUM, OF VIBRATING ENERGY   17.1

STRINGS, QUANTUM VIBRATING ENERGY, ESSENTIAL FOR "BIG BANG"   17.11

STRONG NUCLEAR (COLOR) FORCE BOSONS (GLUONS)   9.4.3   13.4

STRONG NUCLEAR (COLOR) FORCE GLUONS ARE INFINITE IN VALUE AND LONG
        RANGE IN PRINCIPLE 13.4.4

STRONG NUCLEAR (COLOR) FORCE CONSISTS OF EIGHT GLUONS 13.4.1

SUPER-SYMMETRIC PARTICLES 14.1

TEMPERATURE (IN DEGREES)   1.2.23

TENSORS DESCRIBE MOTION IN CURVED SPACETIME 22.11

TETRA- AND PENTA-QUARKS   10.1   11.7.3

TIME DILATION   21.3.2

TIME DILATION AND LENGTH CONTRACTION   21.3   21.3.2

TIMELINE, UNIVERSE, EARTH, LIFE, AND SCIENCE 1.1

TIME MACHINE 21.12

TRANSFER OF HEAT ENERGY BY   2.9   2.9.1   2.9.2   2.9.3

TRAVELING TWIN IS YOUNGER DUE TO ACCELERATION 21.12.3

TRAVELING TWIN IS YOUNGER DUE TO DISTANCE CONTRACTION 21.12.2

TRAVELING TWIN IS YOUNGER DUE TO TIME DILATION 21.12.1

TWIN PARADOX 21.12

 TYPES OF STARS – CREATION OF ELEMENTS IN THE UNIVERSE   6.2.2   6.2.3

UNCERTAINTY PRINCIPLE 12.4.1   12.5

SCIENTIFIC METHOD, UNDERSTANDING NATURE, OUR UNIVERSE, AND BEYOND OUR

UNIVERSE USING   19

UNIVERSE, AGE OF 1.2.14   6.1   6.6

UNIVERSE, BAD NEWS FOR TRAVEL BEYOND   17.12

UNIVERSE, BEGINNING TO END OF   17.10

UNIVERSE, DIAMETER OF   1.2.14   6.1   6.8

UNIVERSE, DIAMETER OF, BEFORE THE BIG BANG 1.2.14 6.1

UNIVERSE, EARTH, LIFE, AND SCIENCE TIMELINE 1.1

UNIVERSE IS EXPANDING AT ACCELERATING RATE 6.5

UNIVERSE, HOW MUCH ENERGY DOES IT TAKE TO MAKE?   17.7

UNIVERSE LOOKS THE SAME IN ALL DIRECTIONS 6.4

UNIVERSE, WILL CONTINUE TO EXPAND DUE TO DARK ENERGY 17.13

UNIVERSES, OTHER KINDS OF   16

VELOCITY, FREQUENCY, AND WAVELENGTH OF LIGHT 5.1

VIRTUAL PARTICLES 12.7 17.2

VIRTUAL PARTICLES, COMBINING IS DETERMINED BY PROBABILITIES   17.4.

VIBRATING STRING CHARACTERISTICS INITIATED BIG BANG WITH LESS PRESSURE,
        TEMPERATURE, AND ENERGY THAN EXPECTED   17.8

VIBRATING STRINGS, OF QUANTUM ENERGY, CREATED BIG BANG 17.5

INDEX (CONTINUED)

WARPS IN SPACETIME, WHY OBJECTS WITH MASS FOLLOW 13.7

WAVE INTERFERENCE   3.9

WAVELENGTH 1.2.9   3.3

WAVELENGTH, VELOCITY, FREQUENCY, AND WAVELENGTH OF LIGHT   3.5.2   5.1

WAVE MOTION AND INTERFERENCE 7

WAVE PHASE OF, IS MEASURED IN 360 DEGREES 3.4

WAVE VELOCITY 3.5   3.5.1   3.5.2

WAVES, AMPLITUDE, FREQUENCY, WAVELENGTH, PERIOD, PHASE, AND VELOCITY 3.3

WAVES CANCEL AND REINFORCE EACH OTHER AND THEMSELVES 3.9

WAVES, COHERENT 3.10

WAVES, COHERENT, INTERFERE WITH EACH OTHER 7.1

WAVES, FUNDAMENTAL FREQUENCY AND HARMONICS OF 3.6

WAVE MOTION AND LIGHT CHARACTERISTICS 3

WAVES, PERIODIC, AND BEAT FREQUENCIES (NOTES) 3.11

WAVES, REFLECTION OF 3.7

WAVES, REFRACTION OF 3.8

WAVES, TRANSVERSE AND LONGITUDINAL 3.2

WAVES TRAVEL OUTWARD FROM THEIR SOURCE, OSCILLATING SINUSOIDAL 3.1

WEAK NUCLEAR FORCE BOSONS 9.4.2

WEAK NUCLEAR FORCE BOSONS CONTROL PARTICLE INTERACTION AND DECAY 13.5

WEAK NUCLEAR FORCE BOSON INTERACTION AND DECAY EQUATIONS 13.5.2

WEIGHT 1.2.3, 2.3.2

WORK 1.2.5, 2.4

ZEEMAN EFFECT 4.0